D1254007

Praise for *Synchronicity*

"*Synchronicity* is a very informative and thought-provoking account of humankind's efforts from antiquity to the present to understand the causal structure of the everyday world and, during the past century, to unite that understanding with the apparently acausal nature of the quantum world of atoms and particles. Paul Halpern writes with remarkable clarity and insight in a very accessible and engaging style."

—David C. Cassidy, author of *Beyond Uncertainty*

"Paul Halpern, a gifted expositor of science, takes the reader on an exciting ride through history, showing how physicists and their antecedent philosophers have sought to understand nature through its connections—through synchronicity."

—Kenneth W. Ford, author of *Building the H Bomb*

"In this beautifully written page-turner, physicist Paul Halpern reveals as much about the secrets of the universe as he does about his own masterful ability to uncover hidden patterns in the history of ideas. With elegance, clarity, and penetrating insight, *Synchronicity* takes its readers on a sweeping journey through a dazzling array of intellectual traditions in search of what connects everything around us and beyond. A superb intellectual achievement."

—Julien Musolino, author of *The Soul Fallacy*

"A roller coaster of a ride that covers more than two thousand years of philosophy and physics, culminating in a fascinating brain-bender of what we have to give up if we want to reconcile ideas like 'cause and effect' with the bizarreness of our fundamental reality. Every possible interpretation, including those of the twentieth century's greatest physicists, dissatisfies us in some way. In Paul Halpern's capable hands, our uncertainties about the quantum Universe transform from frustration to wonder as various ideas are at last put to the test, deepening our appreciation of our mysterious quantum Universe."

—Ethan Siegel, author of *Beyond the Galaxy*

"*Synchronicity* is a sweeping account of humanity's understanding of the nature of causality. With great virtuosity, Paul Halpern weaves together all of the threads of this important story, from the ancient Greeks to modern physics, while entertaining the reader with insightful character studies and colorful anecdotes. A delightful book that anyone interested in the history of ideas will enjoy."

—John Kounios, coauthor of *The Eureka Factor*

"The development of scientific knowledge is often portrayed as some kind of superhighway—as a fast route to complete understanding of the world. This is far from the truth. Real science meanders, it twists and turns, runs this way and then that, and sometimes gets hopelessly lost. *Synchronicity* takes us on a delightful journey along the pathways of real science, and Paul Halpern is a most genial guide."

—Jim Baggott, author of *Quantum Reality*

"From what primeval pattern was the fabric of the universe formed? What determined its warp and its woof? In this wonderfully readable book, which somehow never skimps on the science, Paul Halpern explores the mystery of causality through the ever-intriguing collaboration between Wolfgang Pauli and Carl Jung. Masterfully written, it is at once an inspiring story of how quirks and creativity drive breakthroughs in science and a cautionary tale of how our all-too-human desire to seek meaningful patterns can lead even the brightest minds astray."

—Amanda Gefter, author of *Trespassing on Einstein's Lawn*

SYNCHRONICITY

The Epic Quest to Understand the Quantum Nature of Cause and Effect

PAUL HALPERN, PHD

BASIC BOOKS
NEW YORK

Basic Books
Hachette Book Group
1290 Avenue of the Americas, New York, NY 10104
www.basicbooks.com

Printed in the United States of America
First Edition: August 2020

Published by Basic Books, an imprint of Perseus Books, LLC, a subsidiary of Hachette Book Group, Inc. The Basic Books name and logo is a trademark of the Hachette Book Group.

The Hachette Speakers Bureau provides a wide range of authors for speaking events. To find out more, go to www.hachettespeakersbureau.com or call (866) 376-6591.

The publisher is not responsible for websites (or their content) that are not owned by the publisher.

Print book interior design by Linda Mark.

Library of Congress Cataloging-in-Publication Data

Names: Halpern, Paul, 1961– author.
Title: Synchronicity : the epic quest to understand the quantum nature of cause and effect / Paul Halpern.
Description: First edition. | New York : Basic Books, 2020. | Includes bibliographical references and index.
Identifiers: LCCN 2020001433 | ISBN 9781541673632 (hardcover) | ISBN 9781541673649 (ebook)
Subjects: LCSH: Pauli, Wolfgang, 1900–1958. | Jung, C. G. (Carl Gustav), 1875–1961. | Quantum entanglement. | Simultaneity (Physics) | Causality (Physics) | Physics—Philosophy—History.
Classification: LCC QC174.17.E58 H35 2020 | DDC 539.7/25—dc23
LC record available at https://lccn.loc.gov/2020001433

ISBNs: 978-1-5416-7363-2 (hardcover), 978-1-5416-7364-9 (ebook)

LSC-C

10 9 8 7 6 5 4 3 2 1

Dedicated in honor of my father, Stanley Halpern,
and to the memory of my mother, Bernice Halpern

CONTENTS

INTRODUCTION

Mapping Nature's Connections

> *I cannot seriously believe in [quantum mechanics] because the theory cannot be reconciled with the idea that physics should represent a reality in space and time, free from spooky actions at a distance.*
>
> —Albert Einstein to Max Born, March 3, 1947

OUR QUEST TO UNDERSTAND HOW THE COSMOS IS INTERCONnected begins with light. Light races along nature's fast track. It travels through empty space at a phenomenal rate.

Crossing the vast divide between the Moon and Earth, for example, takes less than a second and a half. Compare that to the roughly three days the astronauts of the Apollo 11–manned lunar mission took for their return in 1969. In other words, a light beam is about two hundred thousand times swifter than that groundbreaking space voyage. No wonder we've learned far more about the vast universe by collecting light with telescopes and other instruments than via space journeys.

Nevertheless, Apollo 11 proved vital for science. On that mission, Neil Armstrong and Buzz Aldrin, the two members of the landing crew, left behind a designated bank of mirrors. The reflectors form a critical component of the Lunar Laser Ranging Experiment. Present-day knowledge of the speed of light is so precise that scientists can now aim a laser pulse toward those (and other) mirrors on the Moon to measure its distance with stunning accuracy. Such tests rely on our absolute certainty that light's velocity in empty space is extremely fast, but not instantaneous. Rather it is finite and constant.

For millennia, our ancestors lacked confidence about light's finite speed. The ancient Greeks debated whether light took any time at all to travel through space. Asserting that sunlight must take some time to traverse the space between the Sun and the Earth, the philosopher Empedocles argued for a finite speed of light. While recognizing Empedocles's line of reasoning, Aristotle rebutted that if light traveled through space we would see its intermediate stages. Rather, it must arrive from the Sun instantaneously. In effect, according to Aristotle's views, the speed of light would be infinite.

It was only by the mid- to late nineteenth century that scientists firmly established light's finite velocity. French researchers Armand Hippolyte Fizeau and Jean-Bernard-Léon Foucault developed two different means of measurement, surpassed in precision by the later techniques of American physicist Albert Michelson. Meanwhile, the theoretical methods of Scottish physicist James Clerk Maxwell proved that light is an electromagnetic wave (disturbance due to the interplay of electric and magnetic forces), possessing a constant, finite speed in empty space.

The speed of light has profound importance. As Albert Einstein emphasized in the special theory of relativity, proposed in 1905, it sets the maximum pace of causal interactions in ordinary space. That is, an effect cannot transpire in less time than it would take for its cause, journeying at light speed, to reach the place where it happens. For example, you couldn't somehow rattle a moon rock remotely faster than a laser could reach it. In general, no transaction involving matter or energy might exceed the speed of light traveling through a vacuum. Moreover, anything with mass, meaning most elementary particles, must travel slower than light. It would take an infinite amount of energy to accelerate a massive, subluminal particle up to light speed, which would clearly be impossible.

Though well established by means of numerous experiments, the speed limit for the transfer of matter and information is unintuitive. Why should natural transactions have absolute bounds? In races, records are meant to be broken. With spaceflight, we aspire to travel faster and faster. Banks reward loyal customers by raising their credit limits—offering them a sense of financial freedom, whether real or illusory. No one likes to be fenced in. We like to breach any frontier. Yet, like a sleepy town hoping to slow the pace of tourists rushing through, special relativity imposes a universal cosmic speed limit.

And why that particular value? Did an early, dynamic process cement the speed of light cap, or was it always unwavering? Could there conceivably be other versions of the universe (or universes parallel to ours) where the speed of light is very different? Could there be any enclaves of reality where it is unlimited? Special relativity assumes that the vacuum speed of light is fixed and finite, but doesn't fully explain why.

In 2011, a startling headline in the prestigious journal *Nature*, "Particles Break Light-speed Limit,"[1] describing a new research claim, jolted the scientific community. Physicists could scarcely believe the news. Given the overwhelming support for the basic precepts of special relativity, including the speed-of-light barrier, many were dubious.

The type of particle in question was the extremely lightweight neutrino. First hypothesized by the brilliant quantum physicist Wolfgang Pauli, the neutrino is nearly (but probably not quite) massless and electrically neutral. Consequently, it rarely interacts with other particles—doing so almost exclusively by means of what is called the "weak interaction": a type of interaction involved in certain kinds of radioactive decay.

Neutrinos are exceedingly common. Nuclear reactions in the Sun generate them constantly. They travel rapidly through space and flood Earth every single moment. However, because they interact so rarely, the vast majority simply pass through. Thus, precisely measuring their speeds is challenging. We couldn't simply bounce them, for instance, as we would photons (particles of light), off lunar reflectors and time their return.

The method used by the OPERA (Oscillation Project with Emulsion-tracking Apparatus) team of physicists was to record the time of flight of neutrinos produced at the LHC (Large Hadron Collider) and detected at the Gran Sasso Laboratory, a special facility housed in a highway tunnel to protect from the interference of other particles. The group reported that the neutrinos completed the journey in about 60 nanoseconds' (billionths of a second) less time than the speed of light would predict. While they published their results only after allegedly ruling out a wide range of possibilities for experimental error, alas, other competing teams couldn't reproduce their supraluminal value. Eventually the supposed finding turned out to be purely a glitch in the timing system. Alas, the so-called faster-than-light neutrinos were merely a mirage.

The OPERA experiment notwithstanding, we cannot assume that any scientific hypothesis will last forever. As sacrosanct as Einsteinian special relativity seems today, scientists might someday find a way to circumvent the speed-of-light barrier. Indeed, general theory of relativity, introduced by Einstein a decade after special relativity, contained an important loophole. If space is curved sufficiently due to the matter and energy within in, it might connect up with itself, potentially enabling faster-than-light connections between two otherwise widely separated points. Einstein and his assistant Nathan Rosen formalized that idea in a 1936 paper. The physicist John Wheeler later dubbed such spatial shortcuts "wormholes." While the notion of wormholes remains purely speculative—no one knows if they could be used to bend causal order or if, alternatively, the laws of physics might somehow prevent that—their existence as theoretical solutions to general relativity raises important questions about how nature is connected.

Our intuitive notions about how the universe is organized don't always match its actual structure. Throughout history, widely held conceptions based on common perceptions have crumbled again and again, from Earth-centered models of the solar system to the idea that space is static. Just when we think we have a firm grasp on reality, something wholly unexpected, such as the discovery in the 1920s that the cosmos is expanding, has shattered our confidence.

Perhaps quantum mechanics, with its enigmatic rules that seem to defy physical expectations, drives home that message most starkly. It shows how elementary particles, without communicating through a medium, can nevertheless coordinate their properties over vast distances. Matching such "entanglement" with the evidence of our senses has been an ongoing struggle, since the development of the notion in the 1920s and 1930s.

Entanglement is not an exchange, but rather a correlation of quantum features. In some cases, it acts faster than purely causal communication (involving a chain of intermediate steps taking place at light speed or less) would permit. It allows for two "pipelines" in nature: the information conduit, that operates at the light-speed limit or less, and quantum correlations, which might manifest themselves immediately upon observation.

For all practical purposes, there is no clash between the two. Physics has learned to encompass both. As quantum theorist Časlav Brukner has remarked, "I do not think that quantum entanglement is in any way in contradiction with general relativity. After all, we have quantum field theory on curved spacetime, which is a perfectly functioning theory."[2]

Nonetheless, over the years many scientists have pondered whether or not a fundamental theory that transcends both relativity and quantum physics might offer a unified explanation of how things are connected in nature—from the microscopic scale to the cosmos as a whole. Rather than sew the relationships of quantum physics onto a general relativistic canvas, a unified field theory would start from simple mathematical threads and weave a fully integrated fabric. The result would be a fully quantized theory of gravitation, along with all other interactions and relationships.

Along that vein, one line of reasoning is to posit locality (the properties of each object are determined by conditions in its immediate neighborhood) and causality as emergent phenomena that while absent from the quantum world on its deepest level, arise naturally from the concerted application of its internal logic. Imagine a pointillist painter dabbing on dots seemingly haphazardly, while her audience watches in amazement as an intricate masterpiece, with patterns and themes that unite the entire canvas, unfolds. Similarly, it is conceivable that a nonlocal, acausal, fundamental reality could develop into a web of causal connections between local entities, including the structures of general relativity.

Alternatively, one might propose that the strange features of the quantum world are simply illusions due to our lack of knowledge. In that case one might assume that the rules of classical physics hold sway and try to model entanglement by means of unseen links set in the background—like a sturdy steel skeleton furtively supporting a gossamer skyscraper.

Enacting such a "hidden variables" strategy without violating the results of countless quantum entanglement experiments turns out to be a tall order. However, some stalwart researchers continue to try.

Such unification efforts date back to the work of Einstein, who was frustrated by what he saw as the incompleteness of quantum mechanics. Decrying entanglement as "spooky action at a distance," Einstein argued for a causal web connecting all processes in the universe. If the property of one object is contingent on that of another, one should be able to demonstrate a domino-like series of causes and effects linking them. For that, he drew on common experience. If a volcano erupts on an island hundreds of miles from your beach house, and sometime later your kitchen starts to shake, you might reasonably deduce that a seismic wave passed from the former to the latter. If the rumbling of a nearby construction project turned out to be the true culprit, that would still be an example of causality. In other words, for any given effect, there must be some chain of causes that produced it.

Moreover, according to Einstein, the physical properties of any object should in principle be completely knowable (presuming perfect instruments) and wholly dependent on conditions in its vicinity—a set of criteria called "local realism." Like a weather vane, anything properly measured should reveal where it is perched, how fast it moves, and what in its surroundings causes that motion. However, numerous experiments have shown that quantum entanglement is real, local realism doesn't fully describe quantum interactions, and that the brilliant physicist's intuition about that subject was incorrect. His commonsense views about how natural events must be linked proved deficient. He was right, though, in identifying the issue as a serious philosophical conundrum that shouldn't simply be swept under the rug.

Our hunches about how things are connected often serve us well. Yet at times they are absolutely wrong—not just in physics, but also in our daily life experiences. When our perceptions are correct, it is a marvel to behold. The power of cognition is an extraordinary tool. Keeping an eye toward the future—by collecting data and using it to shape mental models—is a hallmark of our species. Yet, as in optical illusions, our senses can deceive. As the eighteenth-century Scottish philosopher David Hume emphasized, our belief in causal connections stems from our impressions, but those can mislead. Consequently, to map out the complex skein of connections in physics, taking into account the not-so-intuitive rules of relativity and quantum mechanics, we need to make sure we know how to separate the real from the illusory—true patterns from meaningless coincidences—a task that is not always simple.

Many great thinkers over the ages have conflated valid, testable scientific connections with pseudoscientific analyses. The Pythagoreans introduced important insights about mathematics (famously, a critical theorem involving the sides of right triangles) along with specious numerology (extolling certain numbers). German mathematician Johannes Kepler based an early model of the planets on his intuitive sense of geometric simplicity, before turning to experimental data and realizing his initial hunch was wrong. He also sold horoscopes to earn extra income. Nevertheless, once he applied a more systematic approach to planetary data, his theories were right on target. The great English biologist Alfred Russel Wallace discovered the scientific notion of evolution by natural selection, independently of Darwin, but also embraced pseudoscientific beliefs in the power of spiritualist mediums and the validity of séances. The list of those who found both valid and false connections goes on and on. Even scientists can't always separate the real from the illusory.

Consider the idea of synchronicity: a term coined in 1930 by Swiss psychologist Carl Jung as an "acausal connecting principle." Though he'd attribute the idea to dinner discussions with Einstein about relativity, along with personal analyses of dreams, coincidences, and cultural archetypes, the notion took flight after discussions with Pauli about novel aspects of quantum physics that distinguished it from classical mechanistic determinism. In retrospect, Jung's insights about the need for a new acausal principle in science were brilliant and prescient. Nonetheless his low threshold for accepting anecdotal evidence about "meaningful coincidences" without applying statistical analysis to rule out spurious correlations was a serious failing in his work. Jung trusted his intuitive sense of when things were connected. But in light of the mind's capacity to fabricate false linkages at times, pure intuition on its own is not genuine science.

If one dismisses remote acausal connections altogether, however, as pointed out in a famous 1935 paper by Einstein, Boris Podolsky, and Rosen (often called "EPR"), how do two widely separated, but entangled, particles anticipate which property an observer is about to measure? If, for instance, one of the particle's momentum is measured, instantly revealing the momentum of the other one, how did the second one immediately prepare itself? Did it engage in a kind of "mind reading"? Einstein thought not, and argued for a more complete explanation in which physical values objectively exist before they are measured—even if practical device limitations preclude their measurement.

Coincidentally, around the same time that Einstein dismissed the orthodox description of quantum entanglement as a kind of "mind reading" that had no place in objective science, trained botanist J. B. (Joseph Banks) Rhine argued vehemently that purported psychic's claims of mind reading needed to be scientifically explored. For example, could "psychically gifted" individuals guess the images on cards concealed from them at a rate higher than chance would indicate? To that end, Rhine founded the field of parapsychology.

Rhine's arguments caught the interest of several quantum physicists, including Pauli and his friend Pascual Jordan. While Pauli was generally careful not to broadcast his interest in parapsychology, he became passionately interested in the notion of unseen connections. In many areas of physics, such as in his critiques of Einstein's attempted models for unifying the laws of nature, Pauli was a hardheaded skeptic. Yet, in the case of parapsychology, Pauli was surprisingly very willing to believe, at least for a time. Introduced to Jung when the Swiss psychologist psychoanalyzed him, the two embarked on an exploration of the notion of synchronicity, hoping to establish the reality of an acausal connecting principle.

While Jung and Pauli correctly pointed out that science needed to move beyond expectations of determinism and causality, they became overeager in their quest to find examples of acausal connections in nature. They attempted to draw parallels between quantum entanglement and coincidences in the everyday world, including premonitions in dreams, commonalities in culture (what Jung called "archetypes" and attributed to a "collective unconscious"), and so forth. In making such linkages, they unfortunately conflated a genuine scientific enigma—why both deterministic causality and acausal correlations, involving elements of chance, exist side by side in nature—with unproven pseudoscientific speculations. People—even trained scientists—are not always adept at gauging which connections are genuine and which are spurious. In truth, experimentally confirmed long-distance interactions bear nothing in common with the mere feeling that two events share a hidden linkage. Reproducible results, confirmed by the efforts of numerous teams, are the genuine litmus tests; mere hunches are not enough.

Despite its bizarre aspects, quantum physics is far from being fluffy and vague. On the contrary, within the context of its hybrid framework that includes an odd mixture of chance, correlation, and continuity, it produces extraordinarily precise predictions. These include practical applications, such as MRI (magnetic resonance imaging) used every day in hospitals, and superconducting maglev (magnetic levitation) trains, suspended above their tracks for incredible speeds, being tested in Japan.

A research group based in Vienna, led by innovative physicist Anton Zeilinger, has been doing exciting work for years in the areas of quantum teleportation and cryptography. Making use of quantum entanglement, they have been able to teleport information about the quantum states of photons from one place to another across record distances. Among the team's recent ventures is to send photon state information to the Chinese satellite "Micius" with the goal of exploring how entangled systems might be used in cryptography to create nearly indecipherable codes. Their work shows how acausal connections, such as quantum entanglement, are vital and practical.

While theorists scratch their heads over the meaning of the rules for calculation in quantum physics, experimentalists delight in its bull's-eye measurements. To fathom the entirety of nature, we must learn to reconcile the steely girders of relativity with the pliable, but nonetheless powerful, mesh of the quantum world. Sometimes the same system might have causal and synchronous features.

Consider, for example, the Sun. Its light and heat are generated through nuclear mechanisms that rely on quantum rules, yet are released through space at the rate of causal interactions: the speed of light. While philosophers have contemplated the nature of the Sun's energy for millennia, only in the past century have scientists offered a satisfactory solution that involves a medley of processes.

1

TOUCHING THE HEAVENS

Ancient Views of the Celestial Realm

The Sun, seated in the middle of them, looked at the boy, who was fearful of the strangeness of it all, with eyes that see everything, and said "What reason brings you here?"

—OVID, *Metamorphoses* (translated by Anthony S. Kline)

MAPPING OUT THE WORKINGS OF THE COSMOS HAS BEEN humanity's long-standing quest. The wheels within the wheels of astral motions—from the Moon to the Sun to the starry dome, each as seen from Earth—set our calendars, which, in turn, govern our lives in integral ways. From ancient times until the present, we've tried to find relationships between the behaviors of these bodies—first through speculation and then through science.

Understanding such interactions requires gauging their speeds. A connection involving a delay is fundamentally different than one that is instantaneous. Over the ages, as we've learned about the monumental scope of the universe, fathoming which interactions operate at various rates has become paramount. After all, swiftness is relative to size. Any lagging for a brief interval, presuming the rate stays the same, becomes increasingly significant for longer and longer spans.

To model how a city functions, an engineer would need to understand its networks of transportation and communication. A city restricted to pedestrians would have a wholly different character than one laced with multilane highways—especially in terms of how rapidly products would be delivered from place to place. A community in which mobile phones are banned or restricted would operate at a different pace than a locale in which everyone is carrying a phone at all times.

Similarly, deciphering the dynamics of the web of forces and other interactions in the universe requires a precise comprehension of their operating speeds. We now know that the speed of light in a vacuum serves as an important upper limit for causal interactions between objects in ordinary space. By causal, we mean obeying an order of events in which each effect (something being pulled, for example) is preceded by its cause (the thing doing the tugging).

The ancient Greeks understood the critical importance of light. Applying pure deduction, many philosophers of that era associated it with abstract qualities such as love and goodness, as well as physical properties such as brightness and warmth. Trying to fathom how light was conveyed, they debated whether or not it had a finite speed. Lacking modern instruments and methods, they were unable to resolve that question.

Indeed, because light is so swift, even during the time of the Renaissance, roughly two millennia later, scientists such as Galileo Galilei fared little better in ascertaining its speed. He proposed a method of two observers, separated miles from each other, flashing lanterns in succession and seeing if the timing of the light bursts depended on distance. Though his idea was clever, in practice it wasn't precise enough to distinguish between instant and very slightly delayed (by a tiny fraction of a second) signals. Luckily, thanks to nineteenth-century innovators such as Albert Michelson and continued improvement in techniques and technology, we now know its velocity with great precision.

The speed of light is not just important for astronomy. It has turned out to be a critical component of modern theories of how forces work. Comprehending the forces of nature demands models of how they are conveyed through space. Not all forces involve contact. In fact, two of the four fundamental forces, electromagnetism and gravitation, can act over considerable distances. Do they somehow vault instantly from one point to another, or do they take some time? The electromagnetic interaction, as it turns out, involves the exchange of light. The gravitational interaction, though comprising a different mechanism, happens to occur at the same speed. Consequently, knowledge of the speed of light underpins the study of nature's interactions.

Finally, pinning down a finite speed for light has raised profound questions about the nature of communication and causality. In general, speed limits don't seem very natural. As any hurried driver on a long, empty stretch of motorway would attest, on a day in which traffic police were on strike and nowhere to be seen, temptation to push past the barriers would reign over caution.

If the law of cause and effect is bounded by light-speed influences, as it seems to be, what would happen if that speed limit could somehow be circumvented? Might backward causality be possible? Quantum physics includes coherent states and long-distance correlations that operate faster than a causal chain of events, transpiring at light speed, would seem to permit. How does quantum entanglement, and other remote effects, mesh with the light-speed limit?

In short, the discovery of the finiteness of the speed of light sparked multiple chains of scientific inquiry that have continued until this day. Pure philosophy could not pluck that precious fruit. Rather, it needed to be cultivated and harvested through the development of precise scientific techniques.

Worshipping the Sun

The blazing light shone on our ancient ancestors stemmed from the same sources as the light that shines today. Yet back then, in stark contrast to our modern sense of the extraordinary remoteness of the astral dome, it seemed far more immediate. The ancient Greeks, for example, developed a detailed mythology about the Sun and the heavens that closely connected heavenly doings with terrestrial events.

In some parts of the Greek world, the Sun was worshipped as the god Helios, child of the titans Theia (goddess of sight) and Hyperion (god of celestial light). As Hesiod's *Theogony* relates, Helios's sisters were goddesses Selene (the Moon) and Eos (Dawn). Like siblings sharing a play space, the three immortals took turns dominating the sky.

More insight about the adoration of Helios derives from the Homeric hymns, a compendium of thirty-three poems of unknown authorship, similar in style to Homer's works and likely written starting in the seventh century BC. One such hymn celebrates the Sun god as an exalted driver with a shining golden helmet, steering a quadriga (racing chariot driven by four horses) across the sky:

> [Helios] rides in his chariot, he shines upon men and deathless gods, and piercingly he gazes with his eyes from his golden helmet. Bright rays beam dazzlingly from him, and his bright locks streaming from the temples of his head gracefully enclose his far-seen face: a rich, fine-spun garment glows upon his body and flutters in the wind: and stallions carry him. Then, when he has stayed his golden-yoked chariot and horses, he rests there upon the highest point of heaven, until he marvelously drives them down again through heaven to Ocean.

Personifying the Sun greatly restricted the ancients' ability to study its properties. By purporting that Helios had volition, including the capacity to interact with mortals according to his will and whim, no one could study the body he represented as an actual, steady source of energy. Humans, after all, couldn't fully grapple with the nature of a god's power. Consequently, the road to understanding the Sun in a scientific way, including the process by which its light travels through space, began with the Greek rejection of Sun worship.

It would be in the cultural center of Akragas, near the coast of Sicily, where in the fifth century BC progress would be made in understanding the Sun and its illuminations. By then, although the image of Helios driving the quadriga was still widely known—appearing for example on a gold coin—historians surmise that Sun worship was no longer prevalent. While in Akragas there were prominent temples dedicated to Zeus, Hercules, and other gods, there were none specifically devoted to Helios.

In some parts of the Greek empire, the role of Helios was subsumed by Apollo, a widely worshipped, far more complex deity. A source of harmony, culture, and prophecy, Apollo was far more than just a lightbearer. Curiously, Akragas, unlike other, more central Greek cities such as Delphi, apparently did not have a temple dedicated to Apollo either.[1]

For Empedocles, a learned native of that city, the Sun was a source of philosophical speculation, rather than veneration. He wished to understand the ingredients of reality. The Sun, with its ceaseless fire, seemed an important part of the puzzle.

Dawn in the Valley of the Temples

Dawn comes to Akragas each day in the form of blanched pillars, blistering pavements, and a blinding glow. Far from Mount Olympus, but still part of the ancient Greek dominions, the disk of the Sun makes sure to announce its presence there on its daily rounds. The gleaming temples with their monumental Doric columns reflect an ancient truth. While they purportedly mirror the energy and wisdom of the gods, no one could guess that they actually scatter photons produced inside an unimaginably hot nuclear cauldron, before leaping millions of miles across empty space to reach terrestrial structures such as the temples. Reality is often stranger than myth.

The "Valley of the Temples" at the heart of Akragas is, in truth, situated on a plateau, nestled between a ridge and hills; its location chosen for protection against invaders. In most ancient Greek cities, the orientation of each temple aligns specifically with the direction of the rising Sun during times of ritual importance, such as equinoxes—permitting the greatest illumination of its façade during religious ceremonies. In Akragas, however, the situation is more complicated. With a regular grid pattern of streets, oriented to the plateau's topography, the city is an emblem of functionality. Rather than aligning all the temples according to ritual calendars, at least some of them seem to be arranged for practicality—aligned with the city's lattice, rather than with the Sun's arc.[2] Those alignments further suggest a diminished role for the Sun in worship, offering a greater opening to secular analysis of its properties, including the influential speculations of Empedocles.

Born in Akragas around 492 BC, when the city was less than a century old, Empedocles grew up in a family blessed with great wealth. As with other aristocratic Greek youth of his day, numerous servants waited on him hand and foot. He took to wearing flamboyant clothes, including a flowing purple robe, bronze sandals, and a laurel wreath on his head. The extravagant outfit gave him a regal air, with divine pretentions. Not wanting to be seen as a mere mortal, he presented himself as a mystic and healer. Surprisingly, however, rather than scorning the less fortunate, he did the opposite. Politically, he became a strong advocate of equality and democracy (within the context, that is, of his own hierarchical society that discriminated against women). He worked in his community to pass ordinances guaranteeing equality for free citizens. How could someone profess equality while acting like a holy sage? To paraphrase Walt Whitman, his personal contradictions reflected that he "contained multitudes."

The young Empedocles had a ravenous appetite for poetry and philosophy, ingesting the best works of his day, including the philosophical poem "On Nature" by Parmenides, which had a profound influence on his ideas and style, the natural speculations of Anaxagoras, and the musings of the school of Pythagoras. His readings motivated him to scribe his own meditations about the natural world.

Cosmic Ingredients

As in the case of many pre-Socratic Greek philosophers, much of what we know about the views of Empedocles derives from fragments of his writings and secondhand sources that reference his works. One of his works, "On Nature," directly addresses, and in certain ways rebuts, the monist (single substance) worldview of his mentor Parmenides. It also draws a marked contrast with the numerological views of the Pythagoreans, followers of the philosopher Pythagoras. Parmenides had characterized the cosmos as essentially static—composed of one eternal substance that morphs into various guises but remains fundamentally the same over time. Change, therefore, is a complete illusion. Empedocles, in contrast, argued for a dynamic universe composed of multiple interacting elements.

The Pythagoreans contended that numbers and geometry were the fundamental building blocks of the universe. The integers from one to ten and the regular shapes, such as circles and spheres, had a particular significance as the key components of a hallowed natural order. They ascribed to "one," the "Monad," the property of unity, and "two," the embodiment of divisiveness. In general, odd numbers, connected in their mind with masculinity, brought harmony (the Pythagoreans were an all-male group and thereby biased), and even numbers, linked with femininity, led to clashes of opposites. However, "ten," the "Decad," despite being even, represented the sum of the first four numbers, and thus represented inclusiveness and totality.

One of the most sacred symbols of the Pythagoreans was the Tetractys, a representation of the first ten numbers as an equilateral triangle of points arranged in four rows, with one point on the first row, two points on the second, three points on the third, and four points on the fourth. It wonderfully connects the first four numbers, symbolizing various components of nature, with the cosmic wholeness denoted by the number ten.

Ratios of those first four numbers came into play when the Pythagoreans promoted the idea of harmonious musical scales. Simple ratios of tones, they argued, sounded best. They based their cosmic models, involving concentric spheres of celestial orbits surrounding a "central fire" (not the Sun, but an unseen power source, called "Guard," Zeus's watchtower), on such pleasing combinations of musical notes: dubbed "harmony of the spheres."

The Pythagoreans spoke of eight celestial orbs: the Sun, the Moon, Mercury, Venus, Mars, Jupiter, Saturn, and the dome of the stars. Earth's orbit around the central fire was the ninth sphere. To complete the sacred Decad they also lumped in a tenth body, called the "counter-Earth," which orbits the central fire on the opposite side and thus remains forever invisible.

Mathematics is indeed the language of nature. However, by postulating that "all is number," composed of simple integers and shapes, the

Pythagoreans boxed themselves in to an abstract, unrealistic accounting of the universe's ingredients. They used mathematics to proscribe, rather than describe, the cosmos—leading to severe limitations. For example, the Pythagoreans detested irrational numbers, such as pi or the square root of two, because such abnormalities didn't fit into their scheme. Science depends on embracing all numbers—as tools, rather than ingredients.

Eschewing Pythagorean numerology and musicology, Empedocles advocated more tangible cosmic ingredients. His cosmogony consisted of four main substances: earth, air, fire, and water. He called these "roots," like the roots of a plant. Two opposing fundamental forces, love and strife, acted on those elements to generate nature's dynamo.

Love, according to Empedocles, is the universal force of attraction. It brings like substances together and eventually merges them with dissimilar elements. If allowed to act on its own, however, it would lead to absolute uniformity: a bland, inert mixture of earth, air, fire, and water. While perfect harmony sounds idyllic, it doesn't allow for change, and thereby doesn't permit life.

Luckily, strife serves to counterbalance love, by cajoling elements to separate from each other. Over time, strife drives the various substances increasingly apart until they are as distinctive as the bands of a layer cake. Ultimately, if strife were to prevail, it would divide everything absolutely. In that opposite extreme, life would also be impossible. However, each time strife reaches its upper limit, love kicks in, and the cycle of opposites begins anew. Steered by those disparate forces, the elements mix in various combinations, recycling the stuff of the world over and over again.

Empedocles used an artist's palette analogy to help explain his system. Just as artists mix primary colors together to form secondary hues, to decorate, for instance, a painted vase, nature's force of love brings elements together to create its own masterpieces. With a primitive pointillist perspective, he imagined the varied elements to be tiny dots of material, placed side by side, so finely meshed as to appear to be something new, but were actually a pattern of distinct elemental substances.

By including in his vision the possibility of endless cycles, along with many options for change within the course of each cycle, Empedocles produced a very flexible cosmology. He advanced the study of nature by modeling it with pliable components subject to a variety of interactions. While the elements and forces he listed are far from those scientists consider today to be basic components, the essence of his classification scheme was revolutionary and profound for his time. Look hard, and we see in Empedocles vague premonitions of the current notion of fundamental constituents of matter being quarks and leptons, galvanized by four basic interactions: gravitation, electromagnetism, and the strong and weak nuclear forces.

Empedocles was not just interested in modeling the behavior of inanimate substances. He also ventured into the rudiments of biology, as well as the intersection of those sciences, in his study of the senses. He developed a theory of vision, based on the idea that fire (light) attracts more fire. Vision, he proposed, is an affinity between the fire in one's eyes and the fire in another object.

In what is called the "emission theory of light," Empedocles postulated that the eyes emit beams of light, which make contact with other bodies to illuminate them and perceive them. His theory stood in stark contrast to the "reception theory of light," advanced by the Pythagoreans and other Greek philosophers, which held that the human eye picks up rays transmitted by everything that it observes. Given the dearth of empirical observation in that era, neither camp had the tools to prove its vision of vision. Nevertheless, the philosophical debate between the advocates of the emission and reception theories persisted for years.

In a variety of the reception theory, Democritus, a pre-Socratic Greek philosopher born in 460 BC, proposed each object in the world manufactures unlimited replicas of itself, called eidola, which are transmitted through space and taken in by the body, including the eyes as well as the brain. When soaked in by the eyes, eidola form visual images. When directly absorbed by the brain, eidola offer evocative dreams that allow for premonitions about worldly events. Thus, in his view, seeing and soothsaying are simply different manifestations of how we perceive eidola.

One of the founders of atomism, Democritus believed that everything was made of tiny constituents of various shapes and sizes, which could easily be replicated and released. Therefore, eidola from all parts of the world are all around us, awaiting our notice. Why they arrive exactly in the same order that events transpire, rather than being a jumble of past, present, and future transmissions, Democritus didn't explain.

Philosophers of the ancient world crafted their arguments through logic and elegance. Empirical data were largely absent, except for obvious facts such as that water quenches fire. Therefore, Empedocles and his contemporaries were driven by instinct, rather than experiment. And we've seen how gut feelings can often mislead.

According to some accounts,[3] Empedocles's own demise in 433 BC, at around the age of sixty, may have been a case of fire seeking fire: a blazing personality meeting a scorching demise. As dramatized by Matthew Arnold in his poem "Empedocles on Etna," legend has it that in his final act Empedocles climbed Mt. Etna, the highest European volcano, situated in Sicily. Once he ascended to its rim, he flung himself into its sea of fire, as if to signal his godlike bravery and aspirations of life beyond the grave—attempting to prove, perhaps, the immortality of his own soul. Did Empedocles think the fateful influences that would merge his vital essence with the hellish flames would eventually reverse themselves in a renewed cycle of existence? One might only speculate about the manner and reasons for the death of the great philosopher.

As Arnold imagined Empedocles's final cry:

[T]his heart will glow no more! thou art
A living man no more, Empedocles!
Nothing but a devouring flame of thought—
But a naked, eternally restless mind!

To the elements it came from
Everything will return.
Our bodies to earth,
Our blood to water,
Heat to fire,
Breath to air.
They were well born, they will be well entomb'd![4]

Regardless of the corporeal fate, Empedocles did achieve immortality in his scholarly legacy. Many subsequent philosophers have referred to his writings and ideas, which, along with the works of Pythagoreans and the atomists, had a lasting impact on the shaping of science. That pivotal era was the starting gate in a race over more than two millennia to identify the essential components of nature and how they interact via its fundamental forces.

Among the most influential of the "theories of everything" put forth in the ancient world were presented in the works of Plato, a renowned scholar and teacher, who lived from 429 BC until 347 BC. Founder of the Academy in Athens, he was an avid amalgamator of earlier philosophical views, which he brilliantly shaped into original conceptions of the world, described in writings such as the *Timaeus*.

Following the path of the Pythagoreans, Plato had embraced an idealistic view of the cosmos based on a search for perfection. He proposed that the observed universe, with its obvious flaws, was purely an echo of a harmonious eternal domain. Rather than trying to understand the mundane world by analyzing it directly, he suggested peering beyond its blemishes and trying to fathom its pristine blueprints: the realm of what he called "forms."

A form is the ideal, eternal prototype of all realistic, ephemeral objects in the world. Imagine an immaculate grandfather clock ticking endlessly, free from all sources of friction and resistance, its pendulum swinging beautifully back and forth for all eternity. Compared to a cheap watch, bought in a dollar store, that needs to be reset virtually every day, the majestic clock would be far more representative of time. But even better would be an absolutely perfect timepiece, an archetype of the clock, that didn't even present the possibility of ever missing a beat. That would be the form of "time," from which the best clocks could be crafted and the worst watches graded poorly in comparison. Similarly, the symmetry and elegance of the grandfather clock's pristine exterior could be matched against the form of "beauty," and a child's prodigious admiration for the clock's mechanisms matched against the ideal of "wisdom." The mortal child herself would be a reflection of the idyllic essence of a person—an echo of her perfect "soul."

In short, for every object or quality in the world, the tangible has emerged from the ethereal. Individual human souls themselves have emerged from divine perfection—like a grand, austere cherry tree scattering gorgeous blossoms onto a field below. Those blossoms might be muddied or frayed. Yet any vestige of their beauty that distinguishes them from the base soil offers evidence of their supreme origin.

Such emergence lacks the clockwork precision of causality in the modern sense. There are no obvious, indelible chains that link the realm of forms to the everyday world. Rather, the linkage is an amorphous kind of flow that picks up impurities as it touches coarse reality, like a pristine mountain stream winding its way through secret crevices, slipping past isolated hamlets, darkening as it picks up the needles shed by pine tree groves on its banks, and eventually ceding its waters into a murky municipal basin. Hence, Plato's vision allows for more esoteric modes of connections, such as acausal linkages, associations based on numerology, symmetry, and other mathematical principles, and all manner of supernatural influences. Not surprisingly, in the centuries that followed Plato's passing, it would be subject to a vast range of mystical and occult interpretations.

Plato used a famous thought experiment, the "allegory of the cave," to demonstrate how real life could be an illusory shadow of the realm of forms. He imagined prisoners shacked to an interior wall of a cavern, not too far from its entrance, in which they viewed silhouettes on the opposite wall of people and things that passed by outside—soldiers and their weapons, merchants and their wares, and so forth. If they'd never been free (or had somehow forgotten what the outside world was like) the prisoners might mistake the shadows for actuality. Similarly, our mundane experiences comprise merely an illusory shadow play that bears limited resemblance to omnipresent truths.

Like Pythagoras, Plato was enamored with ideal geometries. The orbits of the planets, Sun, Moon, and stars, he likewise argued, must be circular, at least in the ideal realm. Any perceived deviations in astral behavior must stem from an improper reflection of perfect reality, like a visage in a smudged mirror. One key difference between his model and that of the Pythagoreans is that his vision was geocentric: all orbits centered on Earth, rather than around a central fire.

In the *Timaeus*, Plato presented a curiously Pythagorean take on the elements of Empedocles, connecting them with regular polyhedra (three-dimensional shapes with polygon sides, such as triangles and squares). Such elements behave differently, Plato surmised, because of their unique geometric compositions. Mathematicians note a profound distinction between regular two-dimensional polygons, for which there are an infinite number, and regular polyhedra, for which there are only five types: tetrahedra (four-sided pyramids), cubes, octahedra, dodecahedra, and icosahedra. In other words, those five are the only polyhedra in which all sides are identical and equilateral. The Pythagoreans likely discovered that fact, which the Greek mathematicians Theaetetus and Euclid also described. Nonetheless, given that Plato called special attention to those regular polyhedra, they are usually called the five "Platonic solids."

Nature's Hidden Light

After Plato's death, his Academy in Athens stood for many centuries, even into the fledgling era of the Roman Empire, and his philosophy persisted well beyond that. Throughout the ages, Platonism would resonate in the works of many eminent thinkers. In tandem with the Pythagorean belief in numerology, Plato's focus on forms, rather than the physical world, suggested that nature possesses a hidden code and a transcendental perfection. Platonism, in its various incarnations, thereby challenged savants to try to solve that code.

The Roman-era scholar and biographer Plutarch, born in central Greece around 45–47 AD, engaged in a systematic examination of Plato's writings, with the aim of compiling the ancient philosopher's ideas into a complete description of the universe. He traveled widely, weaving his experiences with many Mediterranean cultures into his studies. For Plutarch, one of the central questions was how the material world encountered the domain of forms, infusing otherwise chaotic, inanimate matter with the essence of the ideal, spiritual realm. His synthesis wove many Pythagorean elements, such as numerical relationships, into Platonic philosophy, along with references to ancient Egyptian symbolism, such as the creation myth involving the divine siblings and lovers Isis and Osiris. Prolific and influential, Plutarch would introduce many future generations to ancient Greek debates about the nature of reality. Plutarch's *Lives*, in particular, would become one of the most influential compendiums of biographies of all time.

Many of Plutarch's ruminations blended what we would call scientific speculation and what we would call mysticism. For example, his treatise on the Moon, *De facie quae in orbe lunae apparet* (Concerning the face which appears on the orb of the Moon), presents a diverse range of ideas about what the Moon is like, from a flat, featureless orb to a sister world to Earth with mountains, valleys, and other features (a view Galileo would prove correct). He presents Empedocles's view that it's a "hail-like congelation of air encompassed by the sphere of fire" and details Aristarchus's calculations of its relative size and distance to Earth. He thoughtfully considers the questions of whether moonlight is reflected sunlight and why lunar eclipses occur more often than solar eclipses. Finally, he fancifully imagines the Moon to be a temporary resting place for the spirits of the dead, before either reincarnation on Earth or passage into some form of afterlife. In short, Plutarch's lunar treatise embodies a crux of ancient and modern views, combining solid observations with supernatural visions.

Historians often characterize Plutarch and other Plato-influenced philosophers of his era as "Middle Platonists." That epithet distinguishes them from the Platonists of antiquity, such as Plato's immediate successors in the Academy. It also separates them from the various mystic strains of Plato-inspired philosophy that followed in the early centuries AD, including Gnosticism, Hermeticism, Manichaeism, and Neoplatonism, as well as medieval occult movements, such as Sufism and Kabbalism.

Briefly, Gnosticism, which includes both Christian and non-Christian strains, refers to attempts, from around the early centuries AD onward, to glean esoteric knowledge of the divine, beyond the scope of traditional religious texts and practices. Following Plato's notion of forms, it purports that there is a pure realm of universal truth that transcends the illusory mundane world. Each of these dual realities was created by a separate divine entity: a greater God, ruler of the perfect realm, and a lesser deity, creator of the material world, including humans with their many flaws. According to Gnostic belief, the ancient Hebrews mistakingly worshipped the more mundane creator, when they should have looked beyond that deity to the more perfect spiritual God. In Christian Gnosticism, Jesus was the conveyor of wisdom about the higher realm; other Gnostic tendencies recognized different messengers of truth.

One of the most famous sets of Gnostic writings were the Nag Hammadi Codices, parchments likely written in the fourth century AD, sealed in jars, and unearthed by a group of peasants in Egypt in December 1945 who were digging for rich loam near the base of the Jabal al-Tarif cliff. It conveyed a gospel considerably different than the standard Church canon. The first set of texts became widely known as the "Jung Codex," after it was bought and smuggled out of Egypt by a Belgian antiques dealer, acquired by Swiss psychologist Carl Jung's Institute in 1952, translated and published, and finally returned to Egypt, where it is now housed in Cairo's Coptic Museum.

Hermeticism, a close cousin of Gnosticism, centered on the persona of Hermes Trimegistus ("Thrice-Greatest Hermes"), a legendary prophet of occult knowledge who purportedly channeled many of the attributes of the ancient Greek god Hermes and the Egyptian god Thoth. It represented a mystic, non-Christian belief system. Manichaeism focused on the dualistic (material versus spiritual) teachings of the devout Iranian sage Mani—born into a Jewish-Christian Gnostic family, but founder of his own religion.

Neoplatonism, a philosophical tendency connected with the writings of Plotinus, Porphyry, and other figures associated with the twilight era of Academy, dismisses the Gnostic idea that the material world is corrupt and the spiritual realm pure, in favor of a more complex relationship between the two. Neoplatonism describes a hierarchal process in which a unified entity, called the Monad, produces a chain of effects that infuse the world of matter with spirit. In detailing how one creates many, its terminology and mechanisms for complexity derive from the Pythagorean idea that the world is constructed via numbers. Transcendence, according to Neoplatonists, involves finding one's way past the mayhem of the convoluted world and reconnecting with the primal unity. The credo traced its roots back to Greek mythology rather than the tenets of Judaism or Christianity. Porphyry, in particular, was sharply critical of Christianity. He found biblical writings inconsistent, and thereby suspect, compared to the well-reasoned discourses of Greek philosophy.

Kabbalism and Sufism represent numerous generations of transcendental thinkers who developed unconventional, mystical interpretations of Judaism and Islam, respectively. Their links to Plato and Pythagoras include decoding sacred texts to look for hidden meaning beyond the explicit writings.

For example, the Kabbalist tradition drew parallels between the holy name of God, the Tetragrammaton with its four Hebrew letters, and the sacred Pythagorean symbol, the Tetractys, with its four rows of points arranged in an equilateral triangle. Other connections with the number four include the four seasons and the four classical elements. For those of a mystic bent, such numerical correspondence had transcendent meaning.

It is interesting to consider the role of light within the context of such mystical belief systems. Rather than simply a physical phenomenon to be measured, it represented divine love and sanctity, and a way to transcend mortal limitations. Gnosticism associated light with sacred knowledge of the true spiritual realm. Manichaeism similarly connected darkness with the material world and lightness with the reign of holy truth. Mani himself was known as the "messenger of light." Likewise, Kabbalists associated light with divine power. In key Kabbalist works, such as the *Zohar* (Hebrew for "light" or "splendor"), God's attributes are described as luminous emanations of unimaginable brilliance.

As the eighteenth century Jewish mystic Israel ben Eliezer, known as the Baal Shem Tov, once described the *Zohar* and its relationship to the *Torah*, the traditional Jewish holy book:

"With the light created by God in the six days of Creation, Adam could see from one end of the world to the other. God hid the light away for the righteous in the hereafter. Where did he hide it? In the Torah. So when I open Zohar, I see the whole world."[5]

In such mystic views, divine light flows freely and instantaneously, without reference to any particular speed. As an all-powerful being, God emanates his lumination without hindrance. Yet, paradoxically He might choose to restrict and contain his power. Some Kabbalistic works envision designated vessels through which divine light is conveyed. Such heavenly conduits are intended to channel and limit divine emanations such that their power does not overwhelm mortal beings. Thus, light acts like a fluid and could well possess a finite rate of flow. Unfortunately, the vessels aren't strong enough to hold the light, and they shatter like smashed test tubes. Such breakage creates disorder in the world. According to some sages, pious behavior, including healing the world through charitable acts, might serve to restore God's original vision.

In general, a common theme of those following in the Platonic tradition was addressing the apparent dichotomy between unlimited divine power and the finiteness of mundane interactions. Some groups, such as the Manichaeists and Gnostics, tried to wall off those two domains; others, such as the Neoplatonists and Kabbalists, tried to bridge them through intermediate structures. Such conduits represent hidden interactions between the eternal and mortal realms that only the most righteous individuals, with pious hearts that beat to timeless rhythms and probing minds that seek out divine wisdom, might begin to fathom.

The Lazy Pace of Sunshine

Plato's legacy, however, was far from confined to mystical movements. Centuries of scholars in western and central Europe came to know him mainly through the works of his most famous student, the eminent (and more pragmatic) philosopher Aristotle, who lived from 384 BC until 322 BC. A prolific expounder of ideas in his own right—emphasizing logical inference and basic observation—Aristotle's interpretation had a lasting impact. His systematic study of the workings of nature, including the various causes of motion, drew from Plato's more abstract conception of how the real and ideal realms are connected, but offered far more specifics on what compelled things to move.

Aristotle broke with Plato in veering toward the realistic. In his cosmic scheme, he adopted a modified version of Empedocles's system of elements. Like Empedocles, he believed that everything on Earth is made of earth, air, fire, and water. However, Aristotle added a fifth element to the mix: quintessence, the etherial stuff of heavenly bodies such as the Sun,

Moon, planets, and stars. The reason celestial bodies were shaped out of quintessence, he surmised, was that it was the lightest material. The element earth was the heaviest, with water, air, and fire having successively lighter weights (but still not as light as quintessence). The heavier a physical object's material, the more grounded it was and likely to sink, rather than to rise. Alternatively, bodies made of quintessence have no reason to plunge toward Earth, and therefore can maintain circular orbits. Hence the Sun and other heavenly orbs, according to Aristotle, each revolve around a static Earth. He summarized his views in the treatise *On the Heavens*, written around 350 BC.

Aristotle's concept of dynamics, though revolutionary for his time, was far more primitive than Newton's laws (proposed roughly two millennia later). Aristotle divided motion into two categories: natural versus forceful. Natural motion involves either the state of rest, rising (for light materials such as fire and air), or falling (for heavy materials such as earth and water). The lighter the element, the more swiftly it rises; the heavier the element, the faster it falls. Combinations of the elements might rise or fall at different speeds depending on their composition. Lacking the concept of inertia, Aristotle surmised that any other form of motion requires a direct impetus. Forcing an object to stray from its natural behavior necessitates a continuous push or pull.

In two of his treatises, *Physics* and *Metaphysics*, Aristotle addressed the issue of causality. His emphasis on finding a cause for every effect helped shape the future of science. We must distinguish, however, Aristotle's definition of causality, which allowed for instantaneous and even backward-in-time relationships between causes and effects, from the modern sense of the word that typically implies a future-directed connection. Specifically, the modern definition of causality is restricted by the limits of communication associated with the speed of light.

Aristotle spoke of four different types of causes, each with distinct mechanisms. He classified causes as either material (what an object is made from), formal (how an object is shaped), efficient (how an object's creator fashions it), or final (an object's ultimate purpose). The fourth type—involving future goals rather than past conditions—is probably further afield from how we generally picture causality. Note, however, that a number of physicists in recent years have addressed the concept of "retrocausality," a past-directed situation where an effect precedes its cause.

Aristotle often gave credit to his predecessors, including Empedocles. While generally an admirer of Empedocles's work, in several areas Aristotle was sharply critical. He dismissed Empedocles's emission theory of vision, arguing that it didn't explain the difference between trying to see during daytime and nighttime. If the eyes produce their own fire, he wondered, why can't we make out images in pitch darkness? Clearly, Aristotle concluded, the eyes are mechanisms for detecting light, not producing it.

Another area of disagreement was Empedocles's theory of the workings of the Sun. Despite having absolutely no idea about the Sun's great distance and colossal source of power, Empedocles managed to reach a supreme insight about how its light traveled to Earth. He argued that it must venture through space to reach us, taking a finite amount of time. Aristotle begged to differ, contending that if so, we'd be able to see it move. In that case, history would prove Aristotle wrong, not Empedocles. As Aristotle wrote in *Sense and Sensibilia*:

> Empedocles . . . says that the light from the sun arrives first in the intervening space before it comes to the eye, or reaches the Earth. This might plausibly seem to be the case. For whatever is moved [in space], is moved from one place to another; hence there must be a corresponding interval of time also in which it is moved from the one place to the other. But any given time is divisible into parts; so that we should assume a time when the sun's ray was not as yet seen, but was still travelling in the middle space.[6]

With only a rudimentary notion of the mechanisms of vision, Aristotle was perplexed by the idea that light could travel through space and yet not be seen in transit. Surely, he pondered, if the light starts at one point and ends at another it must traverse all the steps in between. In that case, why can't we view its complete path, like a burning stream of lit fuel?

Aristotle's critique is surprising. Given that terrestrial light (from fires, lightning bolts, and other sources) seems to reach us without etching its route through the space it traverses, why can't sunlight do the same? Perhaps he was thinking mainly of cases in which fog or haze intervene and reveal the path of terrestrial light. Moreover, sunlight in transit can be seen during the glorious displays of sunrise and sunset, when the atmosphere separates its colors like a prism.

No such "haze" could possibly light up the path of sunlight through deep space, however. The closest such effect is the radiation pressure of the solar wind, the stream of energetic particles released by the Sun. Yet of course the ancients had no way of surmising its existence.

One lesson to be learned from Aristotle's critique of Empedocles is that even the most brilliant philosophers had gaps in their understanding. Aristotle rightly pointed out the flaws in Empedocles's emission theory of vision. Yet curiously, in the rare case Empedocles seemingly reversed himself and spoke about the transmission of light, he turned out to be right. He correctly surmised that the speed of light emanating from the Sun is finite. What its actual value was, he had no idea.

Aristotle's manifold notion of causality, with four distinct classes, allowed for the idea that the Sun could instantaneously flood a human eye with its blazing light, without accounting for the time of transmission through space—about eight minutes, as we now know. If the ultimate point of light is seeing, then solar illumination's "cause" could simply be to fulfill that purpose, according to Aristotle's fourth category. Therefore, according to that logic, its transit time would be immaterial. Fortunately, later thinkers would follow in Empedocles's footsteps, consider the possibility of a finite speed of light, and explore how to measure it.

Worlds in Motion

To progress from the primitive notion of the Sun as a god driving a chariot, to that of a light-emitting orb was a tremendous step. Yet the solar visions advanced by Empedocles and Aristotle almost completely lacked specifics. Neither philosopher had the slightest inkling about the size, content, and dynamo of the Sun. (As mentioned, the mechanisms of the Sun's power would not be known until the nuclear age in the mid-twentieth century.)

Aristotle's geocentric model of the cosmos, in which the Sun, the stars, and the observed planets (five, aside from Earth, were known at the time) revolve around the Earth in perfectly circular orbits, proved extremely influential. Based on Plato's notion that circles are the ideal shape, it was aesthetically satisfying. Yet it failed to explain several key features of the nocturnal sky. Notably, it lacked an explanation for retrograde motion, the strange, temporary reversals in direction that regularly occur in the motions of the planets.

The ancients were well aware of retrograde motion, but pinned it at first on the planets' volition. The word "planet" stems from the Greek word *planetes*, meaning "wanderers." Mercury, Venus, Mars, Jupiter, and Saturn simply had a penchant for sometimes taking a break from their regular strolls through the sky and wandering backward instead. However, Aristotle asserted that motion needed to have a physical cause. He favored natural, rather than anthropomorphic, explanations that offered consistency instead of makeshift justifications. By that standard, his geocentric, circular model of the solar system fell short.

Another situation that Aristotle couldn't explain is why planets sometimes appear brighter or dimmer, indicating that they're sometimes closer to us and other times farther away. If they had fixed circular orbits, why didn't they traverse the sky uniformly?

In an attempt to rectify Aristotle's model, in the third century BC, the Greek philosopher Apollonius of Perga, known as "The Great Geometer," proposed the concepts of eccentric circular orbits, coupled with epicycles. Eccentric means revolving in an orbit not quite centered on Earth, but shifted by an amount called an "eccenter." Apollonius used the slight difference to explain why the planets' orbits seem to vary, leading to greater prominence for different planets at various times. Epicycles are circles within circles, like the twirling of a Ferris wheel's cabins while the giant hub itself spins around. Each planet's epicycle centers on a point on the "deferent," or main cycle. In the case of a Ferris wheel, such twirling might make the cabins occasionally appear to go backward while the big wheel spins forward. Similarly, for planets, the combination of small epicycles and large deferents allows them to seem to move backward in the sky during certain intervals.

In the following century, Hipparchus of Rhodes made detailed astronomical observations, including an extensive catalog of stars and constellations. In studying the motions of planets, he began with Aristotle's geocentric system and similarly resorted to eccenters and epicycles to correct the discrepancies between perfect circular orbits and what he observed. Innovatively, he employed an eccenter in the Sun's orbit to explain why, if it has constant speed, the seasons—as measured from equinox to solstice, or the converse—have unequal length. Because the center of the Sun's circular orbit is slightly displaced from Earth, Hipparchus surmised, it looks like it is traveling a bit faster or slower during various times of the year. His model was accurate enough to make credible astronomical predictions about the Sun and other bodies.

The Greek astronomer Claudius Ptolemy of Alexandria, who lived in the second century AD, drew upon Hipparchus's observations, Babylonian astronomical records, and his own sky-watching, to develop the most detailed astronomical system of its era—with even greater prognosticative power than Hipparchus's scheme. Once again, it was based on Aristotle's geocentric model, amended with epicycles and an eccenter, along with yet another tweak called an equant to make it even more predictive. An equant is a point precisely on the other side of Earth from the eccenter point. While the eccenter constitutes the central point of each deferent, the equant constitutes a unique vantage point in which each epicycle twirls at a constant angular rate. The result was a convoluted, but predictive, means of explaining why the Sun and planets seem to move at variable rates—even backward—at different times of the year without abandoning the sacrosanct idea of circular orbits.

Ptolemy fine-tuned his celestial vision to make forecasts accurate enough to match the sky data and presented his findings in an influential treatise called *The Almagest*, considered for centuries the definitive book about astronomy. It matches observational data about the solar system well enough that mechanical planetariums have made use of his methods (modeling epicycles as gears within gears) for informative sky shows.

Ptolemy's model, though vexingly elaborate, became an astronomical canon for many centuries with the *Almagest* serving as its bible. Indeed, as Christianity spread through Europe, it (and other geocentric models in general) became favored by clerics, in part because the Earth's central role seemed consistent with Old Testament allusions to the Sun's suddenly standing still (Joshua 10:13, Habbakuk 3:11) or moving backward (2 Kings 20:11). For similar reasons, the early Islamic world also embraced geocentric systems such as Ptolemy's. Throughout the Middle Ages, in Europe, the Middle East, and North Africa, astronomy remained fixated on the Earth's centrality.

Not that heliocentric systems were actively suppressed. The notion of medieval clergy brutally repressing advocates of Sun-centered systems is simply the stuff of legend. Rather, the literature expressing heliocentric beliefs was sparse and obscure, especially in the face of the dominance of Aristotle's works and the *Almagest*. In antiquity, Aristarchus of Samos (circa 310 BC–230 BC) was known to have advocated a cosmological system with the Sun as a "central fire," around which all the planets, including Earth, revolve in concentric circles. However, his model was rudimentary, lacked predictive power (because of the dearth of astronomical data, and the fact that planets' orbits are not truly circular, but actually elliptical), and paled in influence compared to the sway of notable geocentric advocates. When it came to prognostic heft matched against a bevy of astronomical data (for its time), Ptolemy reigned supreme until the Renaissance.

Nonetheless, because of the shortcomings of Ptolemy's scheme, rivals would eventually emerge. In addition to its complexity and, of course, its inaccurate placement of Earth near the center of the cosmos, another major weakness is that it lacked an explanation for why the celestial bodies followed the paths they took in the first place. The Pythagoreans had argued that the cosmos was somehow tuned to celestial musical notes in a "harmony of the spheres." Empedocles maintained that it was propelled by the dueling forces of love and strife. For Aristotle, it was the lightness of quintessence, or ether, filling the astral bodies that propelled them high above the Earth. However, aside from matching data, Ptolemy offered little explanation for exactly what impetus drove the deferents and epicycles.

Epitome of the Almagest, an influential summary of Ptolemy's work written by the German mathematician Regiomontanus (Johannes Müller

Aristotle, Ptolemy, and Copernicus, depicted from left to right. Original frontispiece from Galileo's *Dialogues on the Great World Systems*. Credit: AIP Emilio Segrè Visual Archives.

von Königsberg) and published in 1496, alluded to the model's complexity, and offered suggestions for simplification, including replacing circular orbits with spheres. That work helped inspire noted Polish astronomer Nicolaus Copernicus to advocate the heliocentric view in his seminal 1543 opus, *De revolutionibus orbium coelestium* (*On the Revolutions of the Celestial Spheres*). A more sophisticated scheme than Aristarchus's (and likely advanced independently), it boldly suggested that the Earth is rotating about its own axis, producing the illusion that the Sun proceeds through the sky in daytime and the constellations do the same at night. The planets (the five known at the time, along with Earth) revolve around the Sun, and the Moon uniquely orbits Earth. After the publication of Copernicus's revolutionary treatise, Ptolemy's influence persisted, but the voice of doubters began to grow stronger.

Yet another shortcoming of Ptolemy's model was that it didn't include accurate dimensions of each orbit. To hone his model of the cosmos, he had tried but failed to estimate the distances of the Sun, the planets, and the stars. The lack of such measurements hampered his ability to offer a credible scheme. Without knowing the basic dimensions of the solar system, let alone the colossal distances to other stars, astronomy would not be able to progress.

In antiquity, the only celestial distance that was known with reasonable accuracy (within 20 percent error) was that of the Earth to the Moon. One reason the Moon's distance was known so well has to do with a fortuitous circumstance. From Earth's vantage point, its disk in the sky matches closely in size with that of the Sun. The near-match enables total solar eclipses, in which the Moon completely occludes the Sun, to occur with regularity over different parts of the Earth.

Hipparchus employed his geometric skills, along with an understanding of an optical phenomenon called parallax, to measure the Moon's distance during the solar eclipse that likely took place on March 14, 190 BC. Parallax is the apparent shifting of an object when seen from two different vantage points. Because that shifting depends on the observed object's proximity (for a given set of observation points, the closer the object, the greater the shift), it offers a valuable yardstick to assess the distances to nearby astronomical bodies.

A simple experiment, involving both of your eyes and a finger on one of your hands, illustrates how parallax works. Hold up the finger about six inches, or so, in front of your nose. Close one eye, looking at the finger with the other, open eye. Notice its position relative to a fixed background, such as a picture on the wall, if indoors, or a tree, if outdoors. Now close that eye and open the other one. Notice how the finger's position, compared to the fixed background, has changed. As you gaze at it, one eye at a time, you'll see that it seems to move significantly back and forth. Next, try the same experiment with your finger twelve inches in front of your nose instead. You'll see that it seems to move a lot less as you alternate which eye you look at it with. Finally, hold that finger a random distance away. Through geometry, and the amount of shifting from the first two measurements, you could then deduce how far away it is the third time. You could check your parallax estimate with a tape measure.

For remote objects, parallax requires a substantial distance between the two observation points. During the eclipse, Hipparchus picked two locales separated by roughly a thousand miles. The first was the Hellespont (now known as the Dardanelles), a strait (through modern Turkey) that connects the Aegean Sea with the Sea of Marmara. There, the eclipse was total; the Sun was completely blotted out. The second was Alexandria, Egypt (then part of the Greek domains), which experienced a partial eclipse. By Hipparchus's reckoning, during the eclipse's peak over that city, the Moon covered four-fifths of the Sun. Given that the full disk of the Sun subtends about one-half of a degree in the sky, the remaining fifth translated into about one-tenth of a degree. That shift in angle was the lunar parallax, which he could then use to calculate the Moon's radial distance relative to the radius of the Earth. He concluded that the Moon was roughly 71 Earth radii away from its center. The correct value is about 60 radii. While he was wrong, at least he was in the right ballpark.

Why couldn't Hipparchus and other astronomers of that day apply the same methods to estimate the distances to the Sun and other stars? In the case of the Sun, during the daytime there is no fixed reference point against which to measure its shifting by parallax. For the stars, only nearby stellar bodies, such as Proxima Centauri, respond well to the parallax method, providing that the two observation points are on the opposite sides of Earth's orbit around the Sun. Unless one is patient enough to wait months, such stellar parallax is not obvious. (Later, those refuting the Copernican belief in Earth's rotation offered the lack of a noticeable stellar paradox as evidence for their false conception.)

The ancients not only didn't know the speed of light, they weren't even unified on the question of whether or not it had a finite speed. The difference in perspective between Empedocles, who had argued for its finiteness, and Aristotle, who had made a case for its instantaneity, could not readily be resolved by the methods available at the time.

The answer to the question of the finiteness of the speed of light would prove integral to resolving a related conundrum: Could forces and other interactions act remotely, and if so would they take time to do so? To understand how the Sun steers the planets in their orbits, science would need to embrace a universal force of gravitation that somehow links celestial bodies over vast distances.

About two millennia after the classical age of Greece, the contributions of three extraordinary scientists, Tycho Brahe, Johannes Kepler, and Galileo Galilei, would help lay the groundwork for Isaac Newton's mathematical theory of planetary motion, including the notion of universal gravitation. The astronomical data of Tycho (as he is known), as interpreted by Kepler, would elucidate the simple rules for the elliptical orbits of planets. Around the same time, Galileo would craft and apply the first astronomical telescope toward the study of celestial objects, showing they are worlds like Earth, some with moons of their own. Then, some decades later, Newton would develop calculus and apply it toward the findings of his predecessors, brilliantly explaining them with three laws of motion and the law of gravitational attraction.

None of those thinkers, however, would be able to deduce the speed of such interactions between bodies, including the light that shines from the Sun and reflects from planets' surfaces, and the gravitation that helps keep their orbits stable. Galileo and other thinkers would have bold ideas about measuring the speed of light, but frustratingly would lack the technology to carry out such measurements. Danish astronomer Ole Rømer would offer the first rough estimate, based on data from one of the satellites of Jupiter discovered by Galileo, but it was still not particularly accurate. Nevertheless, the work of those sixteenth- and seventeenth-century scientists would help pave the way toward a modern understanding of how objects interact with each other via luminous radiation, gravitation, and other means—including, eventually, accurately measuring the speed of light.

Still, it remains unknown if the light-speed limit for sending signals is mutable under any conceivable circumstances. Modern physics addresses extreme situations, such as highly energetic particles and exceptionally strong gravitation. Could such conditions engender ways of circumventing the light-speed limit? Truly the dilemmas that vexed the ancient Greeks still resound today.

2

BY JUPITER, LIGHT LAGS!

Nature . . . is inexorable and immutable; she never transgresses the laws imposed upon her, or cares a whit whether her abstruse reasons and methods of operation are understandable to men.

—Galileo Galilei, Letter to the Grand Duchess Christina

In my opinion, science is only one among many human faculties. Religion and mysticism have nothing to do with science and are not to be regarded as pseudoscience. They are independent subjects of interest.[1]

—Freeman Dyson (remarks to the author regarding the term "pseudoscience")

To resolve how Nature is stitched together, science needed to resolve the question of whether its swiftest interactions, such as the passage of light through space, are fundamentally instantaneous or, alternatively, finite but extremely fast. Ancient philosophers couldn't solve that puzzle through pure reason. Nor could they rely on the judgments of their senses. Sophisticated methods and instrumentation would be required.

No one has an intuitive feeling about the speed of light. When sun glare or a camera flash dazzle us, we often react before we even realize what happened. We never sense that light has taken some time moving through empty space or through the air. Light can travel thousands of miles in far less time than it takes for our brains to process visual information gathered by our eyes. Therefore, gauging whether light speed is instantaneous, or simply extremely fast, has never been obvious.

Yet without a firm value for light speed, science could not have progressed. Establishing the grand extent of the universe required a deep understanding of how light travels from celestial bodies to us. Without that knowledge, the ancients could only speculate about its scope. Unable to glean the true distances to the stars, they also had no ready way of knowing those bodies' gargantuan sizes, often comparable to or even larger than the Sun. If the stars were close, they could well be twinkling crystals embedded in the velvet sky. But, in truth, aside from the Sun, the closest is so far away that even at light speed it would take years to reach.

Wealth and Wisdom

Hardheaded experimental methods advanced significantly during the European Renaissance. On the Italian peninsula, far north of where Empedocles once pondered light's properties, support for scientific inquiry blossomed during the times of the Medicis, a prominent wealthy family that funded and supported many creative endeavors. True, the genuine investigations were muddled with pseudoscientific disciplines such as alchemy and astrology. The fledgling "scientific community" (in today's parlance) was not yet confident enough to distinguish genuine, reproducible data from suspicious, one-off results, untestable forecasts, and spurious correlations. Therefore, it made little distinction between alchemy and chemistry, astrology and astronomy. Nevertheless, with the development of powerful new techniques and instrumentation, authentic scientific inquiry, from the modern perspective, began to emerge.

The culture of Renaissance Tuscany under the Medicis, centered on the city of Florence, harked back in many ways to ancient Greece. Beautiful artwork, crafted by extraordinary painters and sculptors, such as Sandro Botticelli, Donato di Niccolò di Betto Bardi (better known as Donatello), Michelangelo Buonarroti, and Leonardo da Vinci (who was born in Tuscany but spent much of his adult life elsewhere) flourished under their patronage. The flat, austere paintings of medieval times, commissioned by the Church and mainly restricted to religious motifs, gave way to far more vibrant pieces more faithfully depicting nature and human anatomy. That sprouting of realism developed in tandem with a renewed interest in scientific observation, notably seen in Leonardo's notebooks that included lifelike sketches of the human body.

Spurring the newfound creativity in artistic endeavor and natural studies was an even deeper connection with the ancient Greek world: a revival of interest in its rich philosophical tradition. Unlike the medieval Church doctrine that centered on ways in which Aristotle's writings and other Greek text meshed with the Christian canon, discourse during the Renaissance spurred a fresh look at unorthodox interpretations. Such divergent views included twists on some of the more mystical ideas of Pythagoras, Plato, and other Greek philosophers, interpreted in light of the teachings of the major monotheistic religions.

As the Renaissance progressed, conservative Church authorities would begin to view the revival of such heterodox beliefs with greater consternation. Gutenberg's invention of the printing press in the 1440s would allow for the rapid dissemination of a spectrum of ideas, including unconventional opinions. By the sixteenth century, the Protestant Reformation, championed by Martin Luther, Huldrych Zwingli, John Calvin, and others, would compel the Church to assume an increasingly defensive posture. All the while, agencies such as the Inquisition sought to limit heretical works and dissent through various types of threats, such as banishment, excommunication, incarceration, and—in rare cases (such as that of Giordano Bruno in 1600)—execution. However, the reintroduced currents of Greek natural philosophy, streams of thought that would ultimately feed a torrent of scientific discovery, could not be halted.

The Medici family, who mixed spectacular wealth and power with a penchant for arcane wisdom, became captivated by the secular and occult knowledge conveyed by Plato-inspired philosophers of all stripes, from Gnostics and Hermetists to Neoplatonists and Kabbalists. Their funding and interest brought attention to ancient texts and obscure lines of reasoning that might not have been widely known otherwise. While supporting works purporting to describe nature through transcendental explorations, the Medicis also valued genuine experimentation (in laboratories) and celestial observation (eventually through telescopes). Members of the family were keen on learning more about divine light, artistic light, and astronomical light. Their enthusiasm for knowledge in all of its guises ultimately helped fuel an extraordinary age of scientific discovery.

Like ancient Athens, Renaissance Florence benefited from a rich tradition of informed citizenry involved in political discourse. As a mercantile society, it broke resolutely with the rigid feudal structure in which power was concentrated in the nobility. Rather, thriving trade (wool, in particular) and financial opportunities generated ample wealth for those savvy enough to take the reigns. The newly rich had an opportunity and motivation to try to influence civil society.

Giovanni di Bicci de' Medici (1360 AD–1429 AD) possessed such an enterprising spirit. From his uncle and cousins, he acquired the Rome branch of a financial institution and moved it to Florence, establishing it as the Medici Bank in 1397. Under his leadership, it grew into one of the most prominent banks in Italy, handling the account of the papacy (especially after its return to Rome in 1417 under Pope Martin V) among many other notable clients. With his newfound wealth, he began the Medici tradition of supporting the arts, including taking the then-unusual step of having frescoes painted on the walls of his house. He also became active in public life, advocating civil democracy, while serving as an elected official in the bankers' guild. Despite burgeoning wealth, because of his modesty and humble generosity he became a very popular figure—helping establish the Medicis as admired benefactors of Florentine culture.

When Cosimo de' Medici, later dubbed "Cosimo the Elder," inherited the family fortune (along with his brother Lorenzo), he became an even grander patron of the arts, including sculpture, painting, and architecture. While further expanding the Medici enterprise, he became even more integrally connected with Florentine life than his father was. Like a hawk, he had a sharp eye for opportunities to swoop down into civil matters and display his sagacity and strength. Opponents of his favored projects would be frightened away. He was so influential that a rival wealthy family, the Albizzis, banished him from the city in 1433, hoping to curb his power. But the bid failed. Only a year later, he was invited back to Florence and resumed a role as, for all intents and purposes, its ruler. Like his father, he professed egalitarian beliefs, while, behind the scenes, turning his great bounty into political might.

Centuries earlier, the Great Schism had divided the eastern and western branches of the Church—what would eventually become known as the Eastern Orthodox and Roman Catholic faiths—based in Constantinople and Rome, respectively. Religious leaders had attempted repeatedly to reunite the factions and reconcile their distinct theological interpretations. In the interim, the western church itself had split, only to reunite under Pope Martin V, lending further incentive for a grand reunion.

One such effort, the ecumenical Council of Ferrara-Florence, was relocated from Ferrara to Florence in January 1439 because of the plague. Cosimo played a major role in bringing the holy East-West conclave to his city, during which time the eyes of the Christian world were focused on it. Negotiations seemed to go well. Both sides embraced a tentative agreement to unite the eastern and western churches, paving over sticking points such as the role of the Pope, the nature of the Holy Spirit, and the type of bread used for Communion. The pact would soon dissolve, however, and the schism remain.

Heading the eastern delegation was John VIII Palaiologos, one of the last of a long lineage of Byzantine emperors. Fatefully, he brought with him the Byzantine philosopher Georgius Gemistus Plethon, who, though born a Christian, had become passionately interested in Neoplatonism. Well versed in the writings of Plato, Plotinus, and others in their original Greek, Plethon argued persuasively for the translation of such works into Latin and their adoption into western libraries.

Plethon brought copies of his treatise, *On the Difference Between Plato and Aristotle*, which proved absolutely eye-opening for the participants, particularly those raised on a diet of exclusively Aristotle. For generations, those in the Latinate countries literate enough to appreciate Greek philosophy likely viewed Plato mainly as Aristotle's teacher, rather than as an independent thinker in his own right. That is, they saw the two philosophers as interchangeable. The subsequent history of the Academy after the death of Plato was largely unknown. Plethon changed all that, bringing to the citizens of Florence a newfound appreciation of the bonafide Plato. In particular, Cosimo became enamored of the notion of refounding the Academy and rekindling its mission. His reasons were opportunistic as well as idealistic. If the world perceived Florence as the new Athens, his image as its innovative leader would shine. He mentioned the idea to Marsilio Ficino, a brilliant youth whose education he sponsored.

Ficino was shy, short of build, unathletic, and possessing weak constitution. He spent many hours alone—face lodged firmly in books or sitting and brooding about his purpose in life. In using astrology to interpret his own natal chart, he blamed the influence of Saturn at the time of his birth for his gnawing melancholy. Ultimately, he found joy and motivation in his reading—learning about divine love, for instance, in the works of Plato. His initial encounter with Plato likely transpired through the writings of the ancient Roman orator Cicero.[2] Astrologically, Ficino attributed his ability to rise above despair to the presence of Venus in the constellation Libra and Jupiter in Cancer during the critical moments when he had left his mother's womb. Accordingly, he'd step out of his womblike study from time to time and pleasantly engage in conversation with others.

In 1453, Constantinople fell to the Ottoman Turks, and the Byzantine reign came to an end. Because of his presence in the Council, John Argyropoulos, a learned Greek philosopher, had become familiar with Florence. Sensing an opportunity to enlighten its population, he decided to emigrate there in 1456. He became a teacher of Greek language and philosophy, helping map out, for Florentines, hitherto uncharted intellectual territory.

The silver lining rimming the dark cloud of Constantinople's fall was the critical opportunities it offered to Western scholars. They could learn about ancient Greece from noted scholars such as Argyropoulos, while collecting books brought over by displaced monks from abandoned libraries. For Ficino, it gave him a chance to pursue the career that was clearly in the stars for him. He latched onto Argyropoulos and set out to absorb as much as possible of the master's teachings. After becoming fluent in ancient Greek and proficient in the nuances of Greek philosophy, he established himself as one of Italy's leading experts.

When Cosimo finally had the opportunity to establish a new Platonic Academy in Florence, he eagerly appointed Ficino as its director. Ficino dedicated himself to acquiring a vast library of ancient Greek, Neoplatonic, and Alexandrian texts, including the works of Plato himself, and translating them into Latin. Many such works were available because of the fall of Constantinople. A deeply spiritual man, he believed that the arcane knowledge gleaned from those works would enrich Christian thinking and present no threat to its authority. Astrology, alchemy, the occult, and orthodox religion were each means of seeking holy truth, he felt. He also became fascinated by the use of talismen to ward off evil spirits and preserve good health. The more roads to divine wisdom, the more glorious the topography through which pilgrims might wander to God.

Ficino jubilantly shared his enthusiasm with Cosimo, who dabbled in alchemy himself. Cosimo's own experiments pertained to medications and metals. He devised a fever-reducing elixir for any ailment and concocted a recipe of mercury and silver nitrate, likely intended to be used as a salve for infections. He also discovered a method for making gold coins heavier without affecting their appearance. History does not record whether or not he used that trick to aggrandize himself.[3]

The Medicis' extraordinary brew of natural philosophy mixed with alchemy and the occult became even more potent when powerful Milanese noblewoman Caterina Sforza took on Giovanni di Pierfrancesco de' Medici, a great nephew of Cosimo, as her third husband. They wedded secretly in September 1497, after she had become pregnant. Their only son together, Lodovico, who would soon be rechristened Giovanni dalle Bande Nere, was born in April 1498. Her husband died later that year, leaving her to defend herself and her children (from various marriages) from the onslaught of the region by nobleman Cesare Borgia, son of the Spanish-born Pope Alexander VI, both allied with the expansionist French king Louis XII. It was auspicious that Caterina survived. As it turned out, her progeny would form a new branch of Medici leadership that would rule Florence and Tuscany for centuries.

Resolute, savvy, and well educated, Caterina loved alchemy and herbology. She inherited a collection of Cosimo's recipes and put her favorites to good use. In her passion for natural discovery she concocted many potions, including moisturizers, beauty creams, hair dyes, fertility drugs, aphrodisiacs, and other medications of varying efficacy. As a powerful woman, who could be ruthless at times when she needed to protect her interests and her family, historians dubbed her "the Renaissance virago" and "the tigress." Niccolo Machiavelli, who encapsulated the successful strategies of powerful individuals (including Cesare) in *The Prince*, met her in July 1499 and saw a darker side to her herbology. He and others alleged that among her herbal concoctions were poisons, which she purportedly applied to letters to the Pope in an attempt to assassinate him.

Caterina's great-grandson, Francesco I, inherited (in a sense) her alchemic interests. He spent many focused hours conducting experiments in a special private laboratory, called the "studiolo," constructed in the Palazzo Vecchio, one of the family palaces. There, he mixed various chemicals, crafted metal objects, cut crystals, investigated how to make porcelain, and designed fireworks. His passion for materials science was so well known that several artists painted him visiting various craft studios. Among those are *Francesco I Visiting His Glass Works*, by Giovanni Maria Butteri, and *An Alchemist's Laboratory*, by Johannes Stradanus.

Astronomical Revival

While Francesco was not a particularly adept experimentalist, his era was a splendid time for science in general. Decades after the 1543 publication of Copernicus's heliocentric theory, representing the greatest rethinking of the cosmos since Ptolemy, interest in observational astronomy had started to blossom. Many published copies of Copernicus's treatise were saddled with an unsigned note to the reader (later revealed to be written by Lutheran minister Andreas Osiander) explaining that it shouldn't be taken literally; rather, it should be considered just a handy calculational tool. Nonetheless, its revolutionary findings began to spread.

In Denmark, Tycho Brahe had resurrected observational astronomy, which had largely gone fallow since the time of Ptolemy, by gathering an unprecedentedly detailed set of astronomical data. Tycho (who is usually referred to by his first name) had studied the subject at the University of Copenhagen and other European universities, and was aghast how little had been accomplished since antiquity. Orthodox astronomers treated Ptolemy's data as sacrosanct and took extra measurements only as needed to fill in some of the gaps. For Tycho, that wouldn't do. Stubborn and haughty, he vowed to transform the field radically—and he did.

Nobody who met Tycho could forget his strange appearance. In his youthful days, a fellow student challenged him to a duel about his mathematical abilities. In the course of the skirmish, the mocking boy fought hard and sliced off the bridge of his nose. Tycho had it replaced with a metal prosthetic (likely brass[4])—rendering the bottom of his nose shiny in the light and generally unsettling.

In 1572, shortly after completing his studies at the age of twenty-six, Tycho made his first major astronomical discovery: the sighting of a blazing new star that had never before been observed. The stellar finding won him acclaim, particularly in his native land, where King Frederick II presented him with his own observatory, specially constructed on the island of Hven. He toiled there for years, crafting his own instruments and meticulously recording the locations of the planets relative to the starry background. In between observations, he filled his time with lavish meals and copious alcohol, which proved his undoing. His drunken rages and arrogant behavior toward his assistants and other islanders rose to a frenzy. Finally, after the locals made it clear that was no longer wanted, he decided to leave.

Luckily Tycho soon found a new royal appointment, albeit in a different land. The Holy Roman Emperor Rudolf II invited him to become court mathematician and to work and reside in Prague's Benatek Castle. There, he resumed his instrument-making and his extensive collection of pinpoint data about the planets, which he recorded in numerous notebooks. He also began to ponder the notion of replacing the *Almagest* with a more accurate vision of the cosmos. Ptolemy's model seemed far too complicated. Tycho found Copernicus's heliocentric system much simpler, but refused to believe that the "hulking, lazy" Earth, as he put it, could physically move through space.[5] Lacking the mathematical insights to interpret his excellent data correctly, he continued to tinker in vain with various attempts to improve upon Ptolemy's mechanisms. One such unsuccessful hybrid model imagined the Earth as static, with the Sun and Moon moving around it, while the other planets revolved around the Sun.

Meanwhile, in Italy astronomy was waging a comeback as well. After Francesco died in 1587, his brother Ferdinando I succeeded him as Grand Duke of Tuscany. Two years later Ferdinando married the brilliant Christina of Lorraine, a member of the French branch of the Medicis. She was the granddaughter of Catherine de' Medici, Queen of France and supporter of the renowned physician and astrologer Nostradamus.

One of Ferdinando and Christina's smartest moves was ensuring that their eldest son Cosimo II, born in 1590, obtained an excellent education in natural philosophy. She arranged for a brilliant tutor, Galileo Galilei, a gifted scientist with an artistic bent, who taught Cosimo for three years. Galileo brought to the Medici court years of insight about the workings of the world, including the discovery of the regularity of pendulum motion. Soon he would be making monumental discoveries in astronomy, including the invention of the astronomical telescope in 1609. For science and its promotion, it would prove an auspicious connection indeed.

Between the time Tycho collected his extensive data of the motions of the planets with naked eye observations using his own instruments, and Galileo pioneered the astronomical telescope, a strange situation had arisen. Tycho's voluminous records awaited careful interpretation. By sheer chance, Johannes Kepler, a young German mathematician who had been working in Graz, Austria, had a strong incentive to assist Tycho with the hope of taking a peek at his data.

Kepler's key motivation was a hypothesis he had about a deep connection between planetary orbits and the five Platonic solids (regular polyhedra): tetrahedra, cubes, octahedra, dodecahedra, and icosahedra. He believed that only Tycho's results were comprehensive enough to allow him to put his supposition to the test. However, Tycho kept his findings close to his vest, motivating Kepler to find a way to visit.

Unlike Tycho, Kepler believed in the Copernican system of heliocentric, circular orbits, which he had learned about from Michael Mästlin, a gifted professor at the University of Tübingen. The model, Kepler thought, lacked certain key details, however. It didn't justify why there were exactly

Portrait of noted astronomer and mathematician Johannes Kepler (1571–1630); Credit: AIP Emilio Segrè Visual Archives.

six planets (known at the time), no more, no less. It also didn't explain why their orbits were spaced at particular distances away from the Sun. Why, for example, did Saturn and Jupiter fly through space along certain cyclical paths, like racehorses trained to run on particular tracks?

One fateful day—July 19, 1595, as recorded in his diary—one possible solution aroused his imagination, like the strains of a beautiful aria. While teaching a class about the conjunctions (lining up) of planets, particularly the regular intervals during which Jupiter and Saturn seem to be aligned with each other in the sky, he noted something interesting about circular orbits and regular polygons, such as equilateral triangles. On such figures, one might inscribe (draw on the inside) and circumscribe (sketch on the outside) two different circles—the latter larger than the former—in such a way that the ratio between their diameters is a set value. Imagine a car tire with a central hubcap and outer tread, designed for some reason with a large equilateral triangle that hugs the hubcap and touches the tread with its three corners, and you might get the general picture. In a flash, it occurred to him that perhaps the ratio between the sizes of Jupiter's and Saturn's orbits could be explained that way.

From that point on, Kepler became obsessed with finding the geometric puzzle pieces with which he might construct an accurate model of the solar system. His thoughts focused on the five Platonic solids, which could each be encapsulated in a larger sphere, and inscribed with a smaller sphere. Then, the shapes could be embedded, with the inner sphere of each comprising the outer sphere of another, stacked like traditional Russian nesting dolls. In other words, each solid rests between two spheres. The resulting spheres, in principle, could delineate the orbits of the planets, with the innermost representing Mercury and the outermost, Saturn.

Brimming with enthusiasm, Kepler reveled in his numerological discovery that the number of spheres needed to encapsulate an embedded set of the five Platonic solids precisedly matched the number of planets. That could hardly be chance, he thought. With the zeal of a Neoplatonic sage, he attributed the concordance to divine providence. In his own mind, he had become privileged to a deep truth about God's plan.

In the manner typical of the times, Kepler's astronomy was mixed with a heavy dose of astrology. To earn extra income, he crafted horoscopes, delineating in his charts deep connections between celestial motions and human affairs. He delighted in the transcendental visions of the Corpus Hermiticum and "Hymn to the Sun," by the Neoplatonist philosopher Proclus.[6] Love of mysticism ran in his family. His mother was an occultist, specializing in herbal remedies, and was eventually charged with being a witch. Secret influences seemed as real to him as overt effects. Thus, purporting a hidden relationship between geometry and the heavens seemed perfectly reasonable to him.

Passionately, Kepler pursued his Pythagorean fantasy. He crafted toy models of the Platonic solids, stacked them together, and strove to match exactly Ptolemy's planetary data, as reinterpreted by Copernicus. But the recorded planetary motions would not cooperate with his elegant supposition. Nevertheless, in 1596 he published a treatise on the subject, *Mysterium Cosmographicum*, in which he graphically described his scheme. It happened to be one of the first books to endorse the Copernican system, which in an early draft he tried to reconcile with the Bible.

In an attempt to explain the Trinity in Christianity, Kepler assigned different roles to three different regions of his model. God the Father, he associated with the center of his system. The Son embodied the set of spherical surfaces, and the Holy Spirit, the empty regions in between. Thus God could be one thing, and three distinct things at the same time.

Kepler sent a copy of his book to Tycho hoping to entice the Danish astronomer's interest. He also hoped to get a glimpse of Tycho's excellent data, with the aspiration that it would support his hypothesis. Tycho's response was critical. Nonetheless, he invited Kepler to visit his observatory at Benatek.

Kepler was thrilled by the opportunity to visit, and potentially collaborate with, the greatest observational astronomer in the world. As he noted in his diary:

"Tycho possesses the best observations, and thus so-to-speak the material for the building of the new edifice; he also has collaborators and everything he could wish for. He only lacks the architect who would put all this to use according to his own design."[7]

Another pressing reason compelled Kepler to leave Graz and join Tycho in Prague. The Counter-Reformation, a Church-sanctioned reaction to the Protestant Reformation that began in the mid-sixteenth century and lasted throughout the seventeenth, was in full force. Under its auspices, a wave of religious persecution targeting Lutherans (which Kepler considered himself, although he had theological differences with the movement) and other Protestants had swept through vast regions of central Europe, including what is now Austria. Kepler recognized that he'd be freer to practice his faith in Prague, which had thus far escaped such dogmatism.

After arriving at Benatek, though, Kepler was not particularly happy. He and Tycho repelled each other like oil and water. Kepler was humble and serious; Tycho, overbearing and boisterous. Kepler disdained the revelry in the castle, which annoyingly broke his chain of thought and rattled his peace of mind. Above all, he was gravely disappointed that Tycho shared his data only sparingly, worried that Kepler might usurp his authority and publish a competing theory. Even after Kepler officially became Tycho's principal assistant, in August 1601, the situation wasn't much better.

As fate would have it, though, Tycho wouldn't live much longer. In October of that year, he died suddenly of a burst bladder. Emperor Rudolf II appointed Kepler as Tycho's successor, offering him, at long last, access to the planetary data he coveted. (For decades, Tycho's familial heirs maintained ownership of that information, but Kepler inherited the opportunity to peruse it.)

The most useful data, Kepler soon realized, were the records of the celestial positions of Mars, painstakingly noted by Tycho and his associate, fellow Danish astronomer Christian Longomontanus. They had focused on Mars because its deviation from the predictions of Ptolemy, Copernicus, and Tycho himself in his own model, regarding its erratic orbit and pronounced intervals of retrograde motion, seemed more significant than those of the other planets. Kepler came to see the struggle to understand Mars as a battle of wills between Nature and human intellect. As he would remark in the dedication of his 1609 book *Astronomia Nova* (New Astronomy) to the Emperor Rudolf II:

> [Mars] is the mighty victor over human inquisitiveness, who made a mockery of all the stratagems of astronomers, wrecked their tools, defeated their hosts; thus did he keep the secret of his rule safe throughout all past centuries and pursued his course in unrestrained freedom; wherefore the most famous of Latins, the priest of nature Pliny, specially indicted him: Mars is a Star who Defies observation.[8]

Nonetheless Kepler persisted, carefully interpreting the collected data from the Sun's perspective. In plotting out the Martian orbit, Kepler was astounded to discover that it matched an ellipse (oval) more than a perfect circle. Tossing centuries of Pythagorean, Platonic, and Aristotelian prejudice aside, he decided to accept the data's message at face value, rather than quashing it under the weight of ancient misconceptions. He developed a detailed elliptical model of the Martian orbit, in which the Sun was one of the foci (two guiding points of an ellipse). Compared to the prior assumption of circular orbits, his hypothesis, which became known as Kepler's First Law of Planetary Motion, matched not only Mars but all the other planets as well.

In reaching his conclusion, a quintessential early example of the application of the scientific method, Kepler was forced to abandon his Pythagorean fantasy of embedded spheres alternating with each of the five Platonic Solids. Kepler's dead end showed the limits of his intuitive sense of natural order. His brilliant mind had forged a connection that turned out to be a beautiful mirage rather than an actual truth. Luckily, he was flexible enough in his thinking to switch gears and uncover veritable laws of nature.

Astronomia Nova introduced two such revolutionary principles. Along with highlighting Kepler's model of elliptical orbits, it also detailed his Second Law: planets traverse wedges of equal area (slices of the oval-shaped regions enclosed by their orbits) in equal times. Sweeping the planets along their orbits is the Sun's invisible influence. While offering a premonition of Newton's subsequent notion of universal gravitation, Kepler did not speculate about the origin, nature, or properties of that force. His work was mathematically descriptive and predictive, rather than explanatory. (Kepler's Third Law, describing the relationship between the distances of each planet from the Sun and the time it takes to orbit, would be introduced in his 1619 book, *Harmonices Mundi*.)

By the time of *Astronomia Nova*, Kepler had gleaned that his methods for scientific study were radical in nature. He felt that it was his duty to describe the entire scope of his investigation, including his false starts and missteps, along with his bold conclusions. As he notes in the book's Preface: "What matters to me is not merely to impart to the reader what I have to say, but above all to convey to him the reasons, subterfuges, and lucky hazards which led me to my discoveries. When Christopher Columbus, Magelhaen [*sic*], and the Portuguese relate how they went astray on their journeys, we not only forgive them, but would regret to miss their narration because without it the whole, grand entertainment would be lost. Hence I shall not be blamed if, prompted by the same affection for the reader, I follow the same method."

In opposition to Kepler's scientific methods, other thinkers of the time clung to more mystical views. For example, the English physician and philosopher Robert Fludd argued that a closer reading of the Bible, along with analyses of Hermetic and early Christian works in the Neoplatonic vein, offered ideal guides to the cosmos. From such readings, he devised his own model of the solar system, with the Earth and the Sun each serving as centers of their own domains, and a third center associated with God, lined up as a triad. Echoing the Pythagoreans, distances between the celestial spheres would be based on musical harmonies. Fludd's treatise, *Utriusque Cosmi*, was published between 1617 and 1621 with numerous engravings about his celestial scheme. Kepler roundly criticized the work for its disconnection from observed reality.

One thing *Astronomia Nova* lacked was a section detailing the data Tycho collected and Kepler used to justify his theories. The Brahe family still had publication rights for those notes, which would be released decades later. Perhaps that is why, when Kepler sent a copy of his book to Galileo, he couldn't convince the independent-minded Italian thinker to abandon his preference for circular orbits.

Forbidden Planets

Meanwhile in Tuscany, upon the death of Ferdinando I in 1609, his son Cosimo II took over as Grand Duke and named Galileo as court mathematician. Because of his youthful age, Grand Duchess Christina played a major role advising her son and supporting creative minds such as Galileo. That appointment ensured Galileo the great freedom to pursue explorations of science unhindered by the need to teach or assume other responsibilities. Inventing the first telescope used for astronomy (Dutch optician Johann Lippershey had developed a more practical telescope the previous year), he turned his attention to celestial mysteries.

Born in Pisa in 1564, Galileo was in his midforties when he became the world's leading observational astronomer. He had already made many important contributions to fundamental physics, including debunking

Galileo Galilei (1564–1642), renowned scientist and inventor of the astronomical telescope, portrayed reading while holding a compass; Credit: AIP Emilio Segrè Visual Archives.

Aristotle's idea that heavier things fall faster. Rather, Galileo showed that gravitational acceleration is independent of mass. (The Flemish mathematician Simon Stevin had independently reached the same conclusion three years before Galileo did, but his work wasn't as widely known.) An apocryphal story imagines Galileo dropping two objects of different masses from the Leaning Tower of Pisa and noting that they arrived at the same time. His breakthrough suggested that something in space conveyed the action of gravitation to any massive body placed at a particular point, regardless of how light or hefty it was. That conveyor of gravitation would later be called the "gravitational field." It would supply the missing piece to Kepler's puzzle of what actually "swept" the planets along their orbits.

As important as Galileo's contributions were to physics, his astronomical adventures were even more monumental. Supplementing Kepler's insights, his impact on our perception of the cosmos was extraordinary. By observing mountains on the Moon, Galileo resolved long-standing questions posed by Plutarch and others about its topography. By noting the phases of Venus, he established unmistakably that it moved around the Sun, helping confirm the heliocentric hypothesis of Copernicus. The Milky Way, Galileo found, is composed of numerous, remote stars, not just a gaseous cloud. He mapped out sunspots, the rings of Saturn, and the four largest moons of Jupiter, which he dubbed the "Medici stars" in honor of his benefactors. Kepler would later coin the term "satellites" to describe such bodies orbiting planets.

In 1615, Galileo sent a kind of progress report to Christina, in the form of a detailed letter, hoping to tantalize his powerful patron with impressive new astronomical discoveries and thereby help ward off critics. Presenting his findings, he argued vehemently that the Copernican model must be correct and the Earth must rotate once every twenty-four hours. Given the vast distances of the stars, it would be preposterous to imagine them spinning around the Earth so quickly. Noting that certain Bible passages mention the Sun halting or reversing in the sky, he cautioned that scripture offers important moral messages, but is not the final word on science. Nature, he pointed out, is also a work of God and can be read like a book using telescopes and other instruments. Hence, scientific discoveries should be accepted at face value and then used to reinterpret biblical passages, not the other way around. While centuries later, the letter's key message of openness to scientific findings would become an official position of the Church, its orthodox clergy were far from ready for it at the time.

Following the famous rhetorical style of Plato of presenting novel ideas by means of debates between opposing thinkers (real or fictional), Galileo presented his findings to a more general audience in a 1632 treatise, *Dialogue Concerning the Two Chief World Systems*. These two systems were the geocentric model of Aristotle and Ptolemy and the heliocentric model of Copernicus (he didn't include Tycho's hybrid theory, which had won relatively few adherents). For political reasons, with the goal of not rankling Church authorities, such as the infamous Roman Inquisition, he tried to present the debate as a balanced discussion of abstract rhetorical positions, rather than as a call for a new astronomy. The preface emphasized that the book was hypothetical. That guise allowed the book to be approved for publication by local Church censors. The three debaters—Sagredo, who represents the impartial, open-minded perspective, Simplicio, who symbolizes the simple-mindedness of the traditional Aristotelian and Ptolemaic approaches, and Salviati, who echoes Galileo's scientific perspective in support of Copernicus—offer a gamut of views. Galileo dedicated the work to Ferdinando II, the son of Cosimo II, who had since become Grand Duke upon Cosimo II's untimely death in 1621.

There was good reason for Galileo to be cautious. The Roman Inquisition held considerable power. In the years of the Counter-Reformation, it was tasked with evaluating books and forbidding those that deviated from established doctrine. Authors of such forbidden books were often reprimanded, blacklisted, and in certain cases punished or even executed. After banning Copernicus's *De revolutionibus orbium coelestium*, it was on the lookout for advocacy of heliocentric views, which had been deemed contrary to scripture.

In 1600, the Italian theologian Giordano Bruno was burned at the stake in Rome's Campo de Fiori after being sentenced to death by the Inquisition for his heretical views. Among the beliefs he held deemed blasphemous was the notion that there are unlimited stars in the cosmos, each orbited by its own worlds harboring intelligent beings. An ardent Copernican, influenced heavily by the Pythagoreans, Bruno felt that the heliocentric model of the solar system implied that Earth had no special place in the cosmic scheme. Rather, it was just one of many planets, each with its own properties. Consequently, what happens in the solar system might not just stay in the solar system, but rather be a ubiquitous feature of the universe. As the concept of a plurality of planetary systems departs from scripture, Bruno was condemned. Note, though, that Bruno also disputed other Church doctrines. Therefore the extent to which his astronomical teachings comprised the main reason for his execution is a matter of ongoing historical debate.

Alas, Galileo's gambit got him into trouble too. Although his book was purportedly balanced, a closer reading by Church officials of the way he portrayed the three debaters' views revealed his bias toward the Copernican viewpoint and implied a mockery of other positions. In February 1633, the Inquisition called him to Rome to face charges of heresy. Backed into a corner, Galileo's only recourse was to plead guilty and apologize for his allegedly blasphemous views. His treatise was placed on the list of forbidden books and he was sentenced to house arrest (at his villa in Arcetri, close to Florence) for the rest of his life.

The historian of science Alberto Martinez has described why the two were treated differently:

> Galileo tried to prove the Earth moves, but the Roman Inquisition arrested him. Decades earlier, in 1596, the Inquisition had censured that very claim in Giordano Bruno's books. Bruno was eventually burned alive, partly because he claimed innumerably many worlds exist: countless suns surrounded by planets. That was a *heresy*, a crime denounced in Inquisition manuals and Catholic canon law—yet Bruno asserted it in writing and when interrogated, just as he defended Earth's motion. In contradistinction, when Inquisitors confronted Galileo, he denied it—he lied. So they convicted Galileo of "vehement suspicion of heresy," not of obstinately asserting heresies.[9]

Martinez further points out how controversial views about causality contributed to both thinkers' troubles with the Inquisition:

> Roman Inquisitors criticized both Giordano Bruno and Galileo for having "imposed necessity on God": that given a certain cause, something else *must necessarily follow*. Galileo argued that since the tides exist, the Earth must be moving, just as a water sloshes in a moving container. But the Pope bitterly disagreed: he insisted that even if the tides move, God could make the Earth such that it doesn't move.

Galileo held his tongue for several years, fearful of horrendous repercussions. Embittered by his forced isolation, he also faced the trauma of starting to go blind. Frustrated by his inability to convey important scientific ideas, particularly those that he had labored on in his youth, he decided to compose another grand exposition featuring his three fictitious debaters. Rather than astronomy, the discussions revolved around physics, including his rebuttals of Aristotle's misconceptions about motion.

Titled *Dialogues Concerning Two New Sciences,* Galileo tried in vain to get it published in Italy. However, the censors wouldn't allow it. Finally, he had the manuscript sent to Leiden, Holland, where it was published in 1638.

One of the many innovative aspects of the treatise was its novel treatment of the issue of measuring the speed of light. Rather than following the Neoplatonic tradition of treating light as a kind of spectral emanation, Galileo believed it was ripe for physical examination. Clearly, he argued, it moves through space and thereby should be studied like other moving things.

In one key passage about the question of whether light's speed is instant or finite, Sagredo asks: "But of what kind and how great must we consider this speed of light to be? Is it instantaneous or momentary or does it like other motions require time? Can we not decide this by experiment?"

Conveying the Aristotelian position, Simplicio responds: "Everyday experience shows that the propagation of light is instantaneous; for when we see a piece of artillery fired, at great distance, the flash reaches our eyes without lapse of time; but the sound reaches the ear only after a noticeable interval."

Sagredo rejoins with a more neutral approach: "Well, Simplicio, the only thing I am able to infer from this familiar bit of experience is that sound, in reaching our ear, travels more slowly than light; it does not inform me whether the coming of the light is instantaneous or whether, although extremely rapid, it still occupies time. An observation of this kind tells us nothing more than one in which it is claimed that 'As soon as the sun reaches the horizon its light reaches our eyes'; but who will assure me that these rays had not reached this limit earlier than they reached our vision?"

Finally, Salviati, the voice of scientific reason (conveying Galileo's personal viewpoint), chimes in with the idea for a clever experiment involving people, separated by a distance, carrying lanterns flashing luminous signals to each other in the dark. By covering and uncovering the lanterns with their hands, they might control the timing of those flashes and time how long they take to travel. As Salviati describes:

> Let each of two persons take a light contained in a lantern, or other receptacle, such that by the interposition of the hand, the one can shut off or admit the light to the vision of the other. Next let them stand opposite each other at a distance of a few cubits and practice until they acquire such skill in uncovering and occulting their lights that the instant one sees the light of his companion he will uncover his own . . . Having acquired skill at this short distance let the two experimenters, equipped as before, take up positions separated by a distance of two or three miles and let them perform the same experiment at night, noting carefully whether the exposures and occultations occur in the same manner as at short distances; if they do, we may safely conclude that the propagation of light is instantaneous; but if time is required at a distance of three miles which, considering the going of one light and the coming of the other, really amounts to six, then the delay ought to be easily observable. If the experiment is to be made at still greater distances, say eight or ten miles, telescopes may be employed . . .[10]

Salviati concedes that he has tried the lantern experiment only for separations of less than a mile. At that brief distance, he couldn't tell if the light was truly instantaneous or took some time to travel. Presumably, that was Galileo's actual conclusion as well. However, Salviati notes that lightning bolts seem to take a brief instant to flash through the sky, taking some time to spread among storm clouds, suggesting that light does indeed have a finite speed.

As conveyed by his surrogate debater, even the great Galileo was challenged by the question of testing whether nature's connections, such as the passage of light, might transpire instantaneously or do they always take some time. Though extraordinarily resourceful, he lived in an age where instrumentation had not yet caught up to his imagination. His dual lantern proposal was truly innovative. Yet given the swiftness of light, compared to the relatively lethargic pace of human reactions, he had little chance of carrying out his experiment with the required precision.

Galileo died in 1642, but his work had a lasting impact on the way science was performed and perceived. In particular, it paved the way for establishing the priority of scientific observations over textual interpretations. Religious thinkers of all faiths started to concede that it would be deceptive for God to mislead human observers with false spectacles of nature. If the universe centered on Earth, why would the Milky Way be full of stars, capable of being viewed through optical instruments alone? By the time of the Enlightenment (in the late seventeenth and eighteenth centuries), belief in a more liberal interpretation of the Bible, consistent with remarkable new discoveries about the universe, had become the mainstream perspective among educated thinkers.

Although Galileo couldn't pin down the speed of light, remarkably it was a suggestion of his about using Jupiter's moons as "celestial clocks" that led to its discovery. Because those satellites circled the giant planet with precise regularity, he proposed that astronomers with telescopes could compile comprehensive logbooks of such orbits. Those timetables could then be matched against earthly clocks, typically set on the basis of local solar time. Such a comparison between celestial and local time, Galileo suggested, could offer a quick check on one's longitude—potentially useful for navigation.

In 1676, the Danish astronomer Ole Rømer, working at the Paris Observatory, was compiling such an orbital timetable of one of the largest Jovian satellites, Io, one of the four discovered by Galileo in 1610. Rømer was engaged in trying to resolve a mystery. The observatory's founder and director, Giovanni Domenico Cassini, had noticed a curious linkage between the timing of Io's eclipses (when it was hidden by Jupiter's disk) and the proximity of Jupiter to Earth. In particular, the eclipses were spaced out more and more as Jupiter receded from Earth, and less and less as it came closer. Hypothesizing that the effect was due to a finite light speed, which led to a lag when Jupiter was farther away, Rømer successfully predicted when an eclipse of Io would happen. In doing so, he finally resolved the long-standing debate between the positions of Empedocles and Aristotle—light does take time to travel.

Rømer's rough estimate would be honed in 1690, when the Dutch astronomer Christiaan Huygens published the seminal work *Traité de la lumière (A Treatise on Light)*. As presented in his book, Huygens combined Rømer's eclipse data with an estimate of the radius of Earth's orbit to approximate the speed of light. While his estimate turned out to be about three-quarters of the correct value (determined centuries later), it was a milestone calculation. Huygens also established the notion of light as a wave, which he used to explain the phenomena of reflection and refraction (the bending of light when it travels from one substance to another with different properties).

No longer was light a mere abstraction—a metaphor for the warm feelings of love and the shining beacon of truth. While poets and philosophers might imagine it flowing from heaven and instantly filling our hearts and minds, such metaphors don't reveal the full picture. In treating light instead as an object of scientific inquiry, the methods of Galileo, Rømer, and Huygens represented enormous progress. However, it would take major theoretical and experimental advances—the invention of classical mechanics by Newton, of classical electromagnetic theory by Maxwell, and of special light-measuring apparatuses by Fizeau, Foucault, Michelson, Morley, and others—plus a flowering of optics, to finally reveal light's finite speed and other properties that guide the very workings of the universe.

3

ILLUMINATIONS

The Complementary Visions of Newton and Maxwell

Oh! that men indeed were wise,
And would raise their purblind eyes
To the opening mysteries
Scattered around them ever.
Truth should spring from sterile ground,
Beauty beam from all around,
Right should then at last be found
Joining what none may sever.

—James Clerk Maxwell, "Lines written under the conviction that it is not wise to read Mathematics in November after one's fire is out"

On Christmas of 1642, the year Galileo died, Isaac Newton, one of the most accomplished scientists of all time, was born in Lincolnshire, England. (According to the Gregorian calendar used today, his birthdate was January 4, 1643.) He was a brilliant man, but not particularly a friendly man. Throughout his career, he was engaged in intellectual quarrels with numerous scientific rivals, such as Robert Hooke and Gottfried Leibniz. Harried students might blame him for inventing calculus—which Leibniz also independently developed, sparking their rivalry. Newton employed calculus as a tool for understanding the dynamics of the universe. In that, he had a firm advantage over his long line of predecessors, from the ancient Greeks to Renaissance thinkers.

The essence of Greek atomism and related notions is that the complex world is constructed from much simpler building blocks. Whether they be atoms, elements, numbers, or symbols, such components add up to a greater whole. In no case do they produce something lesser than the original parts.

Classical mechanics, the description of natural dynamics developed by Newton, offers a similarly atomist (in the general sense) approach, relying on the notion of physical bodies that act essentially as point particles or sets of point particles. If two such bodies collide, they either bounce off each other or combine. They never cancel each other out and vanish. Such is the solidity of the world of classical particles.

A trip to the beach might reveal, however, a different kind of natural tallying in which two things added might lead to something smaller, and in some cases nothing at all. Two ocean waves, created by separate disturbances, might happen to meet in such a manner that the trough of one cancels out the peak of the other, leading to the phenomenon known as destructive interference. The same effect might be seen by generating waves on a tightly held jump rope, having them reflected back and forth from the two ends, and ending up with patterns of large fluctuations punctuated by immobile spots, called nodes. Similarly a product of destructive interference, the nodes represent places where the combination of waves leads to an absence, not an amplification.

Fundamentally, though, ocean waves or oscillations of a rope represent the amalgamated behavior of myriad tiny particles (molecules of water or fiber, respectively). Therefore, on the microscopic level they can be analyzed by means of Newtonian classical mechanics. The cancellations of wave patterns, though notable on the large scale, don't translate into the vanishing of particles. Rather, they simply represent places where the particles don't vibrate as much.

What about light waves? Huygen's treatise splendidly revealed how reflection, refraction, and many other observed optical effects might be explained through an arithmetic of wave fronts. If one observes waves emanating from a light source, like ripples from a pebble tossed into a pond, one might sketch a "light ray" perpendicular to the direction of their swell. That ray symbolizes the collective behavior of the light waves along a certain direction. Now suppose those waves reach a barrier with a different material—traveling, for example, from air to water. As signified by the direction of the incoming and outgoing light rays associated with the wave fronts arriving at and leaving from that barrier, part of the light might be reflected, bouncing back from the interface, and part might be refracted, passing into the new material at an angle different from the original.

The wave theory of light implies that a beam split into multiple parts and projected onto a screen in such a manner that it might recombine would interfere with itself. At some points on the screen, peaks of one beam might line up with peaks of another to form larger peaks (that is, constructive interference) seen as bright lines. In other places, peaks might match up with valleys (destructive interference) and produce dark lines on the screen. The result—as would be demonstrated by Thomas Young in his double-slit experiment of 1800—would be a zebra-striped pattern of bright and dark fringes.

With his belief that particles are more fundamental than waves, Newton rejected Huygens's hypothesis. Instead he pursued a theory of light as being composed of tiny constituents of varied colors, called corpuscles. Using glass prisms, he divided white light into its spectrum of hues and reassembled them back into white, revealing in his experiment how water droplets after a storm sometimes break up light into different colors and produce rainbows.

While Huygens's wave theory was very successful, Newton did have evidence on his side to support the corpuscle theory. Consider the distinction between sound and light. Sound, clearly conveyed by waves, creeps around corners and bounces off the walls of caves as echoes. Light, in contrast, creates sharp shadows. Aside from transparent and translucent materials such as glass, visible light seems easily blocked by surfaces. Having not lived long enough to witness Young's double slit experiment, Newton would never be convinced that light was wave-like. He remained a firm believer that nature's varied phenomena, including light, consists of bodies in motion of various shapes, sizes, masses, and other properties.

Each was guided by the strict principles of classical mechanics, predicting its acceleration, and thereby its behavior.

It would take the work of the nineteenth-century Scottish physicist James Clerk Maxwell to complement Newton's corpuscle description of light with a fully developed wave description, and the twentieth-century development of quantum mechanics to explain how the particle and wave descriptions were connected. Until then, they were competing visions of light—with Newton, in his eminence, possessing the upper hand.

But optics was far from the only aspect of science Newton explored. A true polymath, he sought to understand the workings of creation. In essence, he wanted to know what made nature tick.

Action at a Distance

Newton's quest to map out the rules of nature was motivated, in part, by his desire to explain Kepler's laws of planetary motion. Why specifically, he wondered, did planets move in elliptical orbits around the Sun (as one of the foci), sweeping equal areas in equal times? Could a simple truth explain such complex behavior?

Reportedly, it was the falling of an apple from a tree that inspired Newton's solution. More significantly, he noted that if the falling of bodies on Earth related to a more general kind of gravitational attraction, the orbits of planets around the Sun and satellites (such as the Moon) around planets could readily be explained. Such a universal gravitational force, he noted, would need to depend directly on the product of the masses of the bodies in question, and inversely on the square of their mutual distance.

Although Newton argued that such a force would act over significant expanses, such as the vast distance between Saturn and the Sun—let alone the enormous gaps between the stars—he did not specify the speed in which such an attraction would operate. For all intents and purposes gravitational pull would act instantaneously. As he didn't sketch any kind of medium by which gravitational interactions would be conducted (what later became called "gravitational fields"), they operated, in a sense, through a kind of instant, remote magic—akin to a magician waving a wand and seeming to instantly transport his assistant from one side of the stage to the other.

Newton himself recognized the incompleteness of his notion of "action at a distance." In a 1693 letter to the theologian Richard Bentley he wrote:

> It is inconceivable that inanimate brute matter should, without the mediation of something else which is not material, operate upon and affect other matter without mutual contact . . . That gravity should be innate, inherent, and essential to matter, so that one body may act upon another at a distance through a vacuum, without the mediation of anything else, by and through which their action and force may be conveyed from one to another, is to me so great an absurdity that I believe no man who has in philosophical matters a competent faculty of thinking can ever fall into it.[1]

In order to replicate Kepler's laws, along with defining the properties of gravitational forces, Newton needed to define a set of dynamic principles governing how things move. In three laws of motion, he detailed the actions of forces on objects. In tandem, Newton's descriptions of gravitation and motion beautifully implied all of Kepler's rules—a relatively basic calculation readily completed by undergraduate physics majors.

Newton's first principle—the law of inertia—has two parts: one related to objects at rest; the other to moving bodies. An object at rest, he proposed, remains at rest unless acted on by a net force. That statement seems intuitive; as "couch potatoes" know, a restful state seems to sustain itself. The second, less obvious, ordinance is that a moving object keeps going at the same speed along a straight line unless acted on by a net force. In other words, no extra force is needed to keep an object moving linearly. A baseball thrown in deep space would theoretically keep going forever in a straight line. (In practice it would eventually be caught up in the gravitational nets of stars and planets.)

Newton's second law determines what happens if there are unbalanced forces on an object: they compel changes in motion. A net force on a body causes it to accelerate by an amount that inversely depends on its mass. The greater its mass, the less it would accelerate under the action of the same force. Expressed mathematically in a simple equation: force equals mass times acceleration.

Newton's third law concerns the reaction force that echoes back during any action from one body to another. It mandates that the action force from the first body to the second is equal in magnitude but opposite in direction to the reaction force.

Why does the state of inertia not require a force? Why do forces cause acceleration, rather than steady motion? How come reaction forces always emerge, but inevitably on the object causing the action? Newton's laws illustrate how our intuitive sense of connections in the natural world is not always valid. Yet it is essential to explaining important natural phenomena such as the Moon's behavior.

It is a delicate combination of inertia and gravitational attraction that keeps the Moon in orbit around Earth. Without gravity, but only inertia, it would keep on going in a straight line. Without inertia, but only gravity, it would plunge toward Earth. Nicely, the two effects combine to steer the Moon in a regular closed orbit. A similar balance permits stable orbits of planets around stars, including Earth around the Sun.

Classical mechanics, the science of motion on the broad scale from human beings to planets as defined by Newton's principles, is refreshingly simple but enormously powerful. It purports that through a simple equation relating the total force on any object to its acceleration—net force equals mass times acceleration—the movements of that body and its interactions on others might in principle be tracked indefinitely. Assuming perfect instrumentation that might register positions, velocities, and forces with absolute precision, the theory is perfectly predictive on a wide range of scales—accurate enough, for example, to launch spacecraft to the moon.

In classical mechanics, assuming all parameters are measured precisely, there is no room for chance on a fundamental level. Randomness arises for large, complex systems with multiple components, such as a spinning roulette wheel. However, even for such a gambling device, knowledge of mechanics might allow for reasonable enough predictions to make sound bets. For example, in the 1970s a group of young scientists that nicknamed themselves the "Eudaemons" (after a Greek term for happiness used by Aristotle) designed tiny computers and transmitters, programmed with the rules of classical mechanics, placed them in their shoes, and went to casinos with the goal of beating the odds in roulette.[2]

Despite the successes of classical mechanics, there are a few major philosophical gaps in the theory. One glaring omission is an explanation of how certain forces, such as electric, magnetic, and gravitational interactions, act over large distances, such as the mutual attraction of the Earth and the Sun. Huygens, for example, deemed the idea of long-range gravitational attraction "absurd."[3] Only in the nineteenth century would the theories of Carl Gauss, Michael Faraday, James Clerk Maxwell, and others explain how intermediaries called "fields" convey forces from point to point.

Another nebulous aspect of classical mechanics is the need to define a fixed framework, called absolute space and absolute time, for which accelerated motion might be compared and measured. Absolute space means an imaginary set of rigid axes—somehow crisscrossing the cosmos like construction girders framing a building—standing as a benchmark against which distances, speeds, and accelerated motions might be measured. Absolute time is a constantly ticking universal clock that sets the pace of all physical processes. Without absolute space and absolute time, the difference between accelerated and inertial (non-accelerated) motion in classical mechanics would be poorly defined.

For example, a child riding a horse on a merry-go-round might erroneously believe she is not revolving (and thereby accelerating, according to classical mechanics), but, rather, think that the rest of the world is spinning around her. (As hard as it is to imagine a child thinking that he or she is the center of the universe, let's suppose that for argument's sake.) Suppose she is holding a toy, accidentally lets it go, and it flies off the carousel. She might think it was a force that flung it, but rather it was pure inertia, or a lack of force. Reference to a fixed frame—absolute space—would be required to show that despite her misconception, she really was accelerating. Earth is accelerating too, but not at the same pace. Therefore, for commonplace situations, such as carnival rides, the ground is considered to be at rest.

Inspired by the critique of the Austrian physicist Ernst Mach, who deemed absolute space and time artificial constructs, Albert Einstein would be motivated in the nascent decades of the twentieth century to banish those constructs. Still, the power of classical mechanics is such that for practical purposes scientists still pretend that these imaginary yardsticks and clocks exist. And they remain grateful to the founder of the theory, whose personality was as compelling, but flawed, as the mechanics he developed.

Laplace's Demon and Spinoza's God

Newton was a religious Christian, not a pure materialist. Therefore, even though his laws of motion, along with his principle of universal gravitation, brilliantly described all manner of mechanical systems, including the clockwork mechanisms of the planets, he left plenty of room in his model of the cosmos for divine intervention. Humans have free will and are created in the image of God. Therefore, he argued, God, who is infinitely powerful, must have unlimited free will, with the ability to override physical principles for the sake of goodness at any time when that decision is justified. Examples of such divine intervention would be the miracles described in scripture. Natural properties that seemed especially favorable, he believed, could well have been the product of God's benign judgment.

He noted features of the solar system that suggested the actions of an intelligent agent. For example, he thought it remarkable—and possibly God's choice—that the planets, unlike the comets, orbited along the same plane, in the same direction. God could well have placed planets in their orbits and set their rules of motion—in effect "winding up" the system before letting it function.

In a letter to Richard Bentley, Newton pointed out the curious contrast between the Sun, which shines, and large planets such as Jupiter and Saturn, which don't. Newton also attributed that distinction to God's volition: "Why there is one body in our system qualified to give light and heat to all the rest, I know no reason, but because the author of the system thought it convenient; and why there is but one body of this kind I know no reason, but because one was sufficient to warm and enlighten all the rest."[4]

In Newton's interpretation of the Bible's chronology, throughout history, divine miracles and other supernatural interventions supplemented the normal workings of nature—much like a factory foreman, after finding a defect, might choose to stop an assembly line that otherwise functions automatically. Otherwise the world would not be such a benevolent place.

Other thinkers of Newton's day, however, began to wonder if the principles of classical mechanics could be used to explain everything in the cosmos, aside, perhaps, from the creation of those laws themselves. That is, once wound up, the universe would run like clockwork. In essence, there would be no supernatural miracles.

Predating, by decades, the formation of those laws, in 1641, the French philosopher René Descartes's highly influential treatise *Meditations on First Philosophy* appeared. In it he introduced the notion, dubbed Cartesian dualism, that the mind and body are composed of two different substances. That separation would make it easier for physical scientists to avoid the question of conscious volition altogether (due to "mind stuff") and consider a fully mechanistic cosmos that might operate completely on its own.

Later in the seventeenth century, the Dutch philosopher Baruch Spinoza equated God with natural perfection. That identification meant that God was absolutely constrained in his choices and had no power to alter what was already the best of all worlds. The cosmos, wound up, could run like a perfect timepiece.

In the eighteenth century, Deism, the concept of a deity that never intervenes in human or cosmic affairs, aside from the creation of the universe and its laws, became a popular belief in Europe and North America. It rejected the supernatural completely in favor of scientific reasoning. Famous Deists included prominent thinkers such as Benjamin Franklin and Thomas Jefferson. Jefferson, though he considered himself a Christian, was so adverse to the idea of supernatural intervention that he put together his own version of the Bible (now called the "Jefferson Bible") by literally cutting out almost all of the references to miracles and rearranging what he believed were important historical accounts and ethical lessons into a supernatural-free treatise.

Newton's laws of motion, which seemed to explain the actions and interactions of all material things, from the rolling and bouncing of horse-drawn carriages to the glowing chariots of the celestial realm, meshed perfectly with the new scientifically based faith. By eschewing Newton's mystical, religious writings and focusing only on the *Principia*, his masterful account of physics, noted thinkers such as the French mathematician Pierre-Simon Laplace began to fashion a strictly mechanistic, deterministic view of the cosmos.

Born in 1747, Laplace made a name for himself by resolving some of the most formidable problems in classical mechanics. For example, he considered the question of erratic behavior in the orbits of Jupiter and Saturn that couldn't be explained simply by the Sun's gravitational influence. Brilliantly, he showed how the two giant planets tug on each other like arguing siblings and thereby disrupt what would otherwise be simpler elliptical behavior. He presented his solution, and the answers to many other conundrums in the classic textbook *Classical Mechanics*, published in five volumes during the early nineteenth century.

Mixing a measure of philosophy with his science, Laplace speculated that with enough information about the complete set of objects and forces in the natural world, Newton's laws could theoretically predict their behavior indefinitely. In a hypothetical scenario now known as "Laplace's demon," he noted that any intelligent beings able to gather complete knowledge of the positions and velocities of all objects in the universe at any point in time, along with all of the forces that act on them, could calculate their subsequent locations and speeds using Newton's equations of motion. From that information they could compute the new sets of forces, recalculate using Newton's laws, and determine their next set of positions and velocities. Repeating such a process again and again indefinitely, they could achieve absolute awareness of the entire future of the bodies in the universe. As Laplace wrote:

> We may regard the present state of the universe as the effect of its past and the cause of its future. An intellect which at a certain moment

> would know all forces that set nature in motion, and all positions of all items of which nature is composed, if this intellect were also vast enough to submit these data to analysis, it would embrace in a single formula the movements of the greatest bodies of the universe and those of the tiniest atom; for such an intellect nothing would be uncertain and the future just like the past would be present before its eyes.[5]

The clockwork view of the cosmos espoused by Laplace and others profoundly influenced philosophical thinking for many decades. It implied that all of cosmic history is a rigid, chain-linked fence of cause and effect, impossible to sever with any instrument including force of will. Indeed it suggested that "free will" itself is a persistent mirage—an illusion spawned by yet-unknown deterministic mechanisms.

A close cousin of Laplace's demon is Maxwell's demon—invoked in the context of a scientific principle called the Second Law of Thermodynamics, also known as the law of non-decreasing entropy. Proposed by the nineteenth-century German physicist Rudolf Clausius, that principle mandates that no mechanical system runs with perfect efficiency; it must always produce waste energy. "Entropy" refers to the portion of energy that is unavailable to perform work. On the fundamental level, it represents the tendency toward bland uniformity—the opposite of uniqueness.

A delicate snowflake perched on the rim of a mug of hot water is unique. A mug of lukewarm water is less so. Therefore, the latter situation has higher entropy than the former. As a natural example of entropy increase, it is easy to imagine the snowflake falling into the mug and slightly lowering the water's temperature. The opposite, though—a fully formed snowflake suddenly emerging from a lukewarm liquid, ceding its energy, and making it warmer—would be virtually inconceivable. Unless, that is, a magical, microscopic agent managed to perform such a trick.

Maxwell was certainly imaginative. He whimsically proposed that a tiny demon could be tasked with distinguishing slower- and faster-moving molecules in a material. Standing at the portal between two chambers, it could propel the former into one and the latter into the other, stopping each from returning to the wrong side. In that manner, it would ensure that one section would become cooler and the other hotter, decreasing the overall entropy of the system (aside from the demon), and subverting the second law.

Reversibility is an emblem of Newtonian classical mechanics. Its deterministic rules run the same backward and forward in time. Irreversibility is the signature of thermodynamics. Maxwell's demon, at first glance, seems to suggest a way of restoring reversibility to physics—offering hope for reconciling clockwork dynamics with the rusty machinery of entropy increase. However, deeper understanding of the situation reveals that even a demon must generate unusable energy through the waste products of his respiration, digestion, and other physiological processes. Hence the overall entropy of the system, including the demon, increases nonetheless.

Swift Currents and Probing Minds

Born in Edinburgh in 1831, Maxwell soon moved with his family to Glenlair, a country estate on the banks of the River Urr in Scotland. As a youth, he was enthralled by the working of nature. He'd often lie quietly and patiently on the grass and observe its sights and sounds—from the drifting of clouds to the twittering of birds. He'd often ask his parents how and why things move. "What's the go of that?" he'd inquire persistently.

Maxwell's precociousness was demonstrated at the age of two and a half, when he was handed a tin plate to play with as a makeshift toy. He quickly put it to use as a solar reflector, bouncing sunlight off its shiny surface. Excitedly, he asked the family servant to find his parents, who ran into the room, wondering what was going on. After reflecting the sunlight toward their already beaming faces, Maxwell triumphantly announced: "It is the sun and I got it in with the tin plate."[6] It was a harbinger of his lifelong fascination with the properties of light, about which he would make enormous contributions to our understanding.

After being tutored at home for a number of years, Maxwell's parents sent him to the Edinburgh Academy. There he spent much time keeping shyly to himself and working on models and diagrams. The other boys were merciless in mocking him for his quiet, studious ways—ironically nicknaming him "Dafty." From there, he went onto Edinburgh University

Scottish physicist James Maxwell (1831–1879), who revealed that light is an electromagnetic wave; Credit: AIP Emilio Segrè Visual Archives.

and Cambridge—where he acquired the mathematical toolkit that he'd apply so brilliantly to probing connections underlying the natural world.

The mid-nineteenth century—corresponding to Maxwell's most productive years—was an age of considerable progress in measuring the speed of light. Inspired in part by Galileo's notion of using flashing lanterns carried by widely separated observers, two formerly collaborating French scientists, Armand Hippolyte Fizeau and Jean-Bernard-Léon Foucault, independently pursued distinct methods for nailing down its value.

Fizeau constructed a special rapidly whirling cogwheel—hundreds of teeth that would block light, alternating with gaps that would let it through. By shining a powerful beam of light through the wheel, letting the rays travel about five miles before bouncing off a mirror, and arranging for the reflected beam to return through the wheel, he found that he could time the wheel's spin so that light transmitted through a certain gap would be blocked by the next tooth when it returned. Combining the rotation rate with the total travel distance (about 10 miles) offered him a speed-of-light estimate within 5 percent of the modern value of

186,282 miles per second. (Marie Alfred Cornu would refine Fizeau's methods decades later and obtain a reading within 0.2 percent of the actual value.)

Foucault's technique used a rotating mirror instead of a cogwheel. If the system was aligned just right, a beam of light would bounce off of it, hit a second, fixed mirror, and return to the source. However, as the first mirror spun out of alignment, the light would bounce, deflect from the second one, and not come back. Wait an instant, and the alignment would be suitable again. Consequently, the spinning mirror's rotation rate would reveal the speed of light. He repeated the experiment several times, until he achieved an accuracy within 0.6 percent of the correct value. He also placed a tube filled with water along the path the beam of light would take, demonstrating conclusively that light slows down in a fluid, as predicted by Huygen's wave theory, but not by Newton's corpuscle theory.

Foucault also confirmed another of Galileo's hypotheses—Earth's rotation—by hanging a giant pendulum from the ceiling of the Pantheon and observing how its swings gradually precessed (shifted in angle) throughout each twenty-four-hour daily cycle. No one could doubt by that point that the rhythm of day and night was set by the Earth's spin about its axis rather than the Sun's passage through the sky.

Meanwhile, Maxwell's own path to understanding the nature of light—theoretically, rather than experimentally—began when he embarked on a systematic study of the properties of electricity and magnetism. His initial interest in those interactions was sparked by the work of the self-taught English scientist Michael Faraday, who mapped out magnetic influences by scattering iron filings around magnets. To Maxwell's eye, Faraday's diagrams of how such filings fan out from north poles of magnets and converge toward south poles resembled the flow of water away from a source (such as a fountain) and toward a drain. Maxwell speculated that kinds of hidden "fluids"—called field lines—emanated from positive electric charges and magnetic north poles, only to be collected, respectively, from negative electric charges and magnetic south poles. Those electric and magnetic fields offered a more natural way of explaining forces, he thought, than Newtonian action at a distance.

With a powerful visual image in mind of how space is filled with electric and magnetic fields—each, in turn, producing electric and magnetic forces on charges—Maxwell mathematically mapped out their behavior. He found that the two types of interactions are profoundly connected. Changing electric flux (field lines through a surface) produces magnetism, just as (according to an experimental discovery by Faraday called magnetic induction) changing magnetic flux generates electricity.

Maxwell put together a series of equations (later simplified by Oliver Heaviside) depicting the deep connections between charges, electric flux and fields, and magnetic fluxes and fields. Solving the equations, he discovered that they predicted three-dimensional oscillation moving through space: electromagnetic waves. Determining the speed of those waves, he calculated that they approximated the known speed of light, as determined by Fizeau, Foucault, and others. His astonishing conclusion was that light is an electromagnetic wave—a pairing of mutually perpendicular electric and magnetic fluctuations—that travels through space at a speed depending on the medium. The maximum speed of light would be through a vacuum—which seemed at the time an improbable extreme.

All corporeal waves, such as ocean waves or sound waves, must oscillate within a material. Why should light waves be an exception? Therefore, many researchers believed, luminous oscillations must wiggle through a substance so thin and amorphous that it had hitherto escaped detection. It was dubbed the "luminiferous ether"—or "ether," for short.

Seeking the Gold Standard

Maxwell's theoretical calculation of the speed of light matched Foucault's best experimental estimate within 0.5 percent, but that still wasn't precise enough for the scientific community to breathe easy. Experimentalists take pride in honing results until they correspond to a given theory as closely as possible—or in some cases don't, and suggest that it might be wrong. Given the importance of light in astronomy, and its profound connection with electromagnetism, measuring its speed with greater precision seemed vital.

Albert Michelson (1852–1931), posing with the apparatus he used to measure the speed of light; Credit: AIP Emilio Segrè Visual Archives.

An unlikely young researcher, Albert Michelson, would take on that challenge. Born in 1852, in a region of Prussia that is now part of Poland, his family fled Europe when he was a young boy and whisked him to the American Wild West. He grew up in the scrappy gold-mining towns of Virginia City, Nevada, and Murphy's Camp, California—unlikely places for experimental brilliance to emerge. (That unusual background—with some fictional embellishments—would render him the subject of a 1962 episode of the gunslinging television series *Bonanza*, titled "Look to the Stars.")

When he reached the age of fourteen, his family moved to San Francisco, and he could finally begin his formal education. Discovering a deep passion for science, he applied to the U.S. Naval Academy in Annapolis, Maryland, where he didn't make the cut at first. The persistent youth hopped on the train to Washington and met with President (and former Civil War General) Ulysses S. Grant, whose personal intervention led to his admission. After graduating from the Academy in 1873, he returned two years later to become an instructor of chemistry and physics.

In 1877, Michelson's supervisor, Lieutenant Commander William Sampson, suggested that Foucault's rotating mirror apparatus would offer a captivating demonstration for students. Michelson agreed, but decided that it needed improvement. He began to think of ways of making it far more precise, including lengthening its baseline (distance between mirrors) from 60 feet to 2,000 feet, using a compressed-air turbine to spin its main mirror at the dizzying rate of 256 revolutions per second, and testing its drumbeat regularity with a tuning fork.[7] Those steps made the system twenty times more precise than Foucault's best effort (and even better than an 1874 estimate by Cornu, based on Fizeau's method)—offering a marksman's targeting of light's speed and awarding Michelson the recognition of scientists worldwide. More locally, the Canadian-American astronomer and mathematician Simon Newcomb was so impressed that he recruited Michelson to work at the Nautical Almanac Office in Washington, DC, part of the United States Naval Observatory. Michelson was also invited to engage in research at several European universities, where he had an opportunity to develop new optical instruments. In a time when scientific discoveries got only limited press coverage, he'd be featured in an 1882 *New York Times* piece about his work that deemed him, at the age of thirty and just starting a long career, "an accomplished scientist."[8]

It was around that time that Michelson began his first academic appointment, at the Case School of Applied Science in Cleveland, Ohio. While in Europe, he had developed a new light-measuring apparatus involving splitting and recombining beams that would later be dubbed the "Michelson interferometer." Once settled back in the United States, he sought a collaborator with whom to perfect his instrumentation and carry out further optical experiments. He found an enthusiastic partner in the chemist Edward Williams Morley, a colleague at nearby Western Reserve University (which would later merge with Case to form Case Western Reserve University).

In Michelson's previous light-speed observations he had seen no effect due to the hypothesized "ether wind" (effect of the Earth's motion through the ether, also known as the "ether drift"). However, classical mechanics mandates that velocities are additive. A boat traveling in the same direction as a swift current would seem faster from the perspective of someone standing on the riverbank than a boat traveling perpendicular to the current or opposite to the current. That additively implied that light traveling *with* the ether wind should move faster than that traveling *perpendicular to* or *against* the ether wind. Yet no one had detected an influence.

Viewing the lack of evidence as a measuring challenge, he and Morley decided in 1887 to carry out a definitive test. They constructed a large interferometer, based on Michelson's earlier design, to break up a beam of light and send each segment along two different paths at right angles to each other. Hence, they comprised distinct routes relative to the direction of the Earth's motion through space (if one direction was parallel, the other would be perpendicular)—and hypothetically through the ether. The partial beams of light would merge together, in an interference process sensitive to slight differences in path length. If the ether wind slowed down light, distinct zebra-striped interference patterns would reflect a discrepancy between speeds of the partial beams. In particular, they wouldn't be lined up exactly. However, Michelson and Morley found a null result: no effect at all of the ether on the speed of light.

Interestingly, Maxwell's equations were completely agnostic about the existence of ether effects. As a baseline, light could well travel through empty space. Any intervening substance would slow light down. Yet there was no fundamental reason it needed to slow down, aside from the perceptual challenge of imagining waves oscillating in sheer nothingness. For those used to the properties of material waves, such as sound waves and ocean waves, the idea that light waves could travel through pure emptiness seemed startling. Consequently, belief in luminiferous ether persisted for many decades after the Michelson–Morley experiment seemed to rule it out.

To justify the lack of findings, in 1892 the Dutch physicist Hendrik Lorentz and the Irish physicist George Francis FitzGerald independently proposed that the ether wind's pressure caused a slight shortening of objects moving along its direction. Therefore, it skewed the Michelson–Morley apparatus in such a way to equalize the path lengths of the perpendicular beams, enabling them to merge in such a way that there was no record of a speed difference. While that proposal, called the "Lorentz–FitzGerald contraction," seemed an incredible coincidence, those believing in ether had little other recourse.

The Last Hiding Place of Ghosts

By the end of the nineteenth century, a large segment of the scientific community anticipated a time when a complete description of the cosmos and consciousness would eradicate all manner of superstitious beliefs: from omens to ghosts, and from evil demons to divine intervention. In theory, any natural phenomenon could be broken down into a precise sequence of causes and effects. Alleged "miracles" either happened for a reason—a previously unknown mechanism such as a natural product that turned out to have pharmaceutical effects—or were pure coincidences born of wishful thinking. "Show me the precise causal mechanisms," physical science seemed to demand, "or abandon your beliefs."

Those renegade scientists who continued to believe in the supernatural looked ardently for gaps in the Newtonian picture. Rather than exclude the world of the spirit—ghosts, telepathy, strikingly odd coincidences, and so forth, they hope to extend science to normalize those strange phenomena. Thus, they hoped to breathe soul into an otherwise purely mechanical world.

Organizations emerged around that time with the aim of promoting scientific investigations of the paranormal and thereby bringing research about the spiritual realm into the mainstream. Among such groups were the London Spiritual Alliance, founded in 1884, and the Society for Psychical Research, founded in 1885. Prominent members of those groups included the British physicist William Crookes, renowned for his vacuum tube design, the writer Arthur Conan Doyle, famed for the Sherlock Holmes detective stories, William James, an early psychologist, and the German physicist Johann Karl Friedrich Zöllner, a leading expert in optical illusions. Given Zöllner's expertise, his embrace of spiritualist mediums, including famously advocating for the American illusionist Henry Slade, who claimed to channel the thoughts of the dead through writing on slates, was truly ironic.

An important source of purported "evidence" for the realm of the spirit was photographs allegedly showing unusual phenomena, such as ghostly images. While such "spirit photography" could readily be faked using camera tricks such as reflections and double exposure, many notables became duped into believing the ghostly images were real. In particular, Conan Doyle argued vehemently for their validity. Wilhelm Roentgen's 1896 discovery of X-rays—revealing hidden features that optical light couldn't illuminate—would lend further ammunition to spiritualists' claims of realms beyond the tangible.

One would think that hardheaded scientists and other serious thinkers would be skeptical about spirit photography, séances, and other forms of supposed supernatural contact. Indeed most were. But a vocal minority relied on their gut feelings and intuitive sense of the beyond, rather than best scientific practices.

To incorporate spiritual phenomena into science, many research-oriented believers sought nooks and crannies for such phenomena to reside. One potential space was the ether—the hypothesized medium through which light was supposed to move. The mathematician Peter Guthrie Tait, a dear friend of Maxwell's, proposed that atoms and other building blocks of matter are composed of knots in the ether. An avid pipe smoker, Tait reportedly reached that conclusion while blowing smoke rings. Just like a cluster of shoelaces tied so tightly that they seem impossible to unravel, such ethereal knots persist, and thus present the illusion of permanence. According to his theory, each type of atom—hydrogen, helium, and so forth—constituted a different way the ether is knotted. In *The Unseen Universe*, a book cowritten with Balfour Stewart and originally published anonymously, he speculated that thoughts, feelings, spiritual beings, and the soul—that is, nonmaterial things—are aspects of other layers of ether that we can't physically detect. Hence the world of the spirit stealthfully coexists with the material realm in a kind of duality.

Along with the ether, another potential gap in the Newtonian picture embraced by renegade late-nineteenth century scientists looking to justify the supernatural was the possibility of higher dimensions. If, for instance, the conventional three dimensions of length, width, and breadth were supplemented by an unseen fourth spatial dimension it could offer a way of explaining purported extrasensory connections, such as telepathy, telekinesis, and channeling the thoughts of the deceased.

Though once a reputable scientist, Zöllner became convinced that Slade had somehow obtained access to the fourth dimension. One of

Slade's tricks, along with slate-writing, was asking participants to hold the ends of a rope knotted in the middle, and then proceeding to untie the knot without their letting go. In his 1878 book, *Transcendental Physics*, Zöllner explained his theory that Slade reached through the fourth dimension to untie knots, converse with the dead, and so forth.

Maxwell found discussions of knots in the ether and portals to the fourth dimension entertaining, but ultimately lacking evidence. That same year, he teased Tait with "A Paradoxical Ode," a poem in the style of Shelley, which begins with the lines:

My soul is an entangled knot,
Upon a liquid vortex wrought
By Intellect, in the Unseen residing,
And thine cloth like a convict sit,
With marlinspike untwisting it,
Only to find its knottiness abiding;
Since all the tools for its untying
In four-dimensioned space are lying.

Slade went on trial in a London court for defrauding wealthy individuals who paid him for séances in which he would purportedly communicate with their deceased relatives and other dear ones. He'd produce such so-called "messages from the dead" on slates that he'd reveal during the sessions. Once skeptics figured out how he performed his tricks (switching slates, for example), those taken in were irate. Newspapers around the world covered the proceedings, which helped serve to polarize those sympathetic to the spiritualist movement from those who believed it was hogwash. Among the latter, many concurrently began to dismiss the idea of unseen connections in science, such as the possibility of hidden, higher dimensions.

For example, Albert Einstein, who was born in 1879, would have a lifelong skepticism for interactions that didn't transpire through a chain of causal connections. He'd initially be dubious about the fourth dimension, until he needed it for his theories, and would remain cynical about quantum entanglement, deeming it "spooky." Perhaps the debate about spiritualism that transpired in his youth profoundly shaped his hard-headed thinking.

Einstein similarly did not believe in ether, deeming it an unnecessary hypothesis. Eschewing ether completely and developing an independent model of how objects move through space took exceptional independence of thought. Few scientists at the turn of the twentieth century were visionary enough to make that leap. Luckily for the progress of physics, Einstein pursued the subject without prejudice, developing his masterful special theory of relativity that banished the ether, while rendering the fourth dimension something as commonplace as time.

4

BARRIERS AND SHORTCUTS

The Mad Landscapes of Relativity and Quantum Mechanics

The beauty and clearness of the dynamical theory, which asserts heat and light to be modes of motion, is at present obscured by two clouds . . . The first came into existence with the undulatory theory of light; . . . it involved the question: How could the earth move through . . . the luminiferous ether? The second is the Maxwell-Boltzmann doctrine regarding the partition of energy.

—Lord Kelvin, "Nineteenth Century Clouds over the Dynamical Theory of Heat and Light" (lecture delivered at the Royal Institution of Great Britain, April 27, 1900)

THE TURN OF THE TWENTIETH CENTURY REPRESENTED A CRITIcal inflection point in the study of how the universe is interconnected. Before that era, science was trending toward the notion of strict determinism, governed by ironclad causal rules. That meant that each natural interaction happened predictably at a certain pace. Assemble particles, map out their forces, and one might anticipate what would transpire indefinitely into the future. Waves, considered to be particle groupings, similarly wouldn't surprise in their behavior. After that era, however, quantum mechanics would reveal the bizarre inner worlds of atoms—including random, seemingly instantaneous jumps from one state to another, acausal, long-distance correlations between particles, and the very weird uncertainty principle that meant one could never completely ascertain the positions and speeds of elementary particles at the same time. Relativity would show how space and time are flip sides of the same coin, while delineating the boundaries of causal connections. Yet, like many political revolutions, practically no one anticipated that the tide was about to turn until it actually happened.

If one were to poll a group of physicists at a New Year's party in 1900, chances are most would agree on a number of general principles underlying the natural world, including causality, continuity, experimental reproducibility, and the universality of physical law spanning all scales from the tiniest to the largest. Newton's laws, showing how forces trigger changes in motion, seemed to mandate a strict succession of cause and effect. The work of Maxwell and others demonstrated how fields conveyed such forces through space in a seamless fashion. Experience showed that while any kind of apparatus or method carried with it a degree of error, repetition of a similar measurement generally led to similar results, within reasonable bounds that shrank over time with improved equipment and techniques. Statistical methods established clear connections between the realm of the minuscule and familiar, larger-scale properties such as temperature and pressure. In short, the canon of physics seemed more or less complete.

Over a few drinks, however, some of the physicists might concede some loose ends, including the dilemma of why the ether wind seemed to have no detectable influence on light speed, and difficulties using simple theoretical models of light waves in explaining the observed properties of thermal radiation emanating from a blackbody (perfect absorber of light) at any given temperature. Lord Kelvin (William Thomson), in an address from that period, characterized those conundrums as dark clouds hanging over an otherwise sunny vista. Ironically, rather than being minor concerns, it would be those two issues that motivated revolutionary scientific developments in the early twentieth century. If the reveling physicists knew what was about to come, they'd either toast with a bubbling champagne or down a stiff whiskey, depending on their philosophical perspective.

It would take Einstein's genius to unravel these riddles. In his development of relativity and early work on quantum physics, his suggestions for transforming the nature of space, time, and matter were revolutionary and profound. On the other hand, his penchant for tearing down older structures only went so far. Throughout his lifetime he ardently supported rigid determinism, albeit with different mechanisms than classical physics. Although quantum relationships, as they developed, would challenge his suppositions, he'd remain stalwart and stubborn in his views.

From an early age, Einstein pondered the intricate web of connections in nature, wondering how everything related to each other. The gift of a compass from his father Hermann, an electrical engineer, set him thinking about what causes magnetic influences. As Einstein later recounted:

> A wonder of this kind I experienced as a child of four or five years when my father showed me a compass. That this needle behaved in such a determined way did not at all fit into the kind of occurrences that could find a place in the unconscious world of concepts (efficacy produced by direct 'touch'). I can still remember—or at least believe I can remember—that this experience made a deep and lasting impression upon me.[1]

As he grew into adulthood, both the radical and conservative aspects of his theories were shaped by his extensive background in philosophy. Heavily influenced by the determinism of Spinoza and Schopenhauer, he had an instinctive disdain for loose ends in science. In the early 1900s, while working as a patent clerk at a government office in Bern, Switzerland, Einstein met with a group of friends nicknamed the "Olympia Academy" to discuss philosophical ideas, including the works of Mach, which stressed analyzing evidence gleaned from the senses—a form of realism. Consequently, Einstein would charge ahead with novel ideas, such as radically new ways of understanding light, for the purpose of resolving the blackbody problem and other dilemmas, but pull on the reins when the consequences of his theories veered too far from the philosophical notions he held most sacred, such as determinism and realism.

Light's Dual Identity

The blackbody radiation problem pictures a box heated to a particular temperature and then allowed to emit light. From our modern perspective, we might imagine a dark container (a jet-black mug of tea with a lid on the top, for example) warmed in a microwave oven and then set on a table. A theoretical principle from the nineteenth century, called the equipartition theorem, mandates that the energy within the box would be doled out equally to each degree of freedom (distinct ways of moving—in light's case, every possible way it might vibrate). In democratic fashion, the energy would be spread among the full range of possible wavelengths (distance between successive peaks) from short to long. Wavelengths, in turn, vary inversely with frequency (the rate of oscillation, which in visible light corresponds to the color). Therefore, the energy would be divided among the complete spectrum of frequencies as well.

The problem is that short-wavelength modes are easier to fit into a given box than long-wavelength vibrations. It is like trying to fit print of certain font sizes across a page—the smaller the font, the more you can squeeze in. Therefore, by allotting equal energy to each degree of freedom, the tiniest units tend to dominate over the largest because there are so many of the former compared to the latter. For light, that means a skew toward short-wavelength, high-frequency radiation. The frequency range immediately higher than the visible part of the spectrum corresponds to ultraviolet light. Hence this problem is sometimes called the "ultraviolet catastrophe."

The light spectrum includes additional modes of vibration beyond the visible and ultraviolet. Invisible radiation with ever-higher frequencies includes X-rays and gamma rays. Light with frequencies lower than the visible includes infrared, microwaves, and radio waves.

Now imagine heating up a cup of tea to 100 degrees Celsius in a microwave oven. After removing the cup and setting it on a table, ordinary experience shows that it would emit mainly invisible thermal radiation in the form of infrared light. An infrared ("night vision") camera would show that clearly. To the touch, the cup would be scalding hot for a time, but not in any sense a radiation hazard.

By way of contrast, nineteenth-century theoretical statistical physics would wrongly predict a ghastly outcome. Because the cup of tea's energy distribution would be skewed toward the high frequency end, removing it from the microwave would result in a blast of hazardous ionizing radiation, from ultraviolet to gamma rays. Such a horrid experience would be no one's cup of tea. Luckily, in truth, no one's cup of tea would deliver such a horrid experience.

In 1900, the German physicist Max Planck found the right way of modeling blackbody radiation. The key was mandating that the energy of each light wave comes in a small finite amount—a "quantum," as he called it, meaning "parcel." That energy allotment depends on the light's frequency, multiplied by a fundamental value now known as Planck's constant. In other words, high-frequency photons (units of light) are more energetic than low-frequency photons; violet, for instance, is warmer than red.

The brilliance of Planck's idea is that it offers a penalty for short-wavelength/high-frequency radiation that helps balance its advantage of being able to cram better into boxes. It is like a tax on petiteness. Each gamma ray photon, though much more compact in terms of wavelength, requires far more energy than, for example, a radio wave photon. By "discriminating" against the former, in favor of the latter, Planck was able to establish the correct theoretical distribution to match what is actually seen.

While Planck introduced the term as a mathematical construct for calculations, Einstein showed in his analysis the photoelectric effect, published in 1905, that quanta are real. The photoelectric effect involves shining a light on a metal, energizing some of the electrons on its surface, and releasing them into space. As Einstein demonstrated, for each metal there is a minimum energy needed to discharge an electron, which corresponds to a minimum frequency. Shine light below that threshold frequency and the photons won't be energetic enough to free the electrons. For that groundbreaking theory, the essence of today's digital cameras, Einstein would receive the 1921 Nobel Prize in Physics (which happened to be awarded in 1922).

The energy-frequency connection is not very intuitive. Before Planck and Einstein, physicists believed that the only factor that influences light's energy content is its brightness. A blast of light of any color, they thought, should have the same energy, all other factors being equal. Yet for a single photon impacting a single electron, causing an energetic leap, its frequency makes all the difference.

The photon model was in some ways a partial return to Newton's notion of corpuscles. By arguing that light comes in discrete packets, Einstein was suggesting that it possessed some of the characteristics of electrons and other elementary particles known at the time. Yet it also clearly had wave-like properties, including a wavelength and frequency.

From antiquity until the turn of the twentieth century, physics had been moving toward increasing clarity and realism. In classical, Newtonian physics, supplemented by Maxwell's theory of electromagnetism, the classical laws of thermodynamics, and so forth, what you see is what you get. Ambiguity and mystery became reserved for either physical phenomena difficult to measure or transcendental questions, such as "what happens to the soul after death" and "what happened before the beginning of time," which those scientists who acknowledged such possibilities had long ceded to theologians.

Yet with Einstein's proof that light has both wave-like and particle-like properties—wriggling through space like an army of Slinkys in the form of photon "wave packets"—he unknowingly set off a seismic wave that would, within decades, shatter the clear-cut physical tenets he learned at university. The abrupt transformation, dubbed by the philosopher Thomas Kuhn a "paradigm shift," would show how scientists' intuitive sense of how things work doesn't always steer them to the right place.

In tandem with his monumental contributions to the nascent quantum idea, Einstein also developed groundbreaking solutions to other key dilemmas in physics, including revising Newtonian mechanics to account for the constancy of light speed. Einstein saw the need to resolve a profound contradiction. While Newton's laws imply that observers could catch up to light waves, like a tugboat trailing a ship, Maxwell's equations ensure the constancy of light's speed for any observer.

Relative Truths

Einstein once imagined running to try to catch a light wave. According to classical theories of motion, including Newton's principles, one could potentially keep pace with light and observe it standing still. How, he wondered, would electromagnetic theory—which seemed to mandate an invariant speed—account for that possibility?

Einstein found that by making space and time malleable constructs, he could ensure that the speed of light would always remain the same, no matter who is measuring it. He rejected the ether and established the invariance of the vacuum speed of light as a sacred principle. The result of his revision would be the groundbreaking special theory of relativity, published in 1905.

Special relativity describes how measurements of distance and time depend on the relative speed of the observer compared to what is being observed. The "special" indicates that the theory is restricted to nonaccelerating systems. In other words, it describes only objects moving at constant velocity.

Some of special relativity's major predictions include time dilation, length contraction, and the equivalence of mass and energy. Each effect is most notable when an object is moving close to the speed of light. Time dilation involves the slowing down of a clock moving in tandem with a high-speed object (inside a spaceship, for instance) as observed by an outsider not traveling at that rate (a terrestrial astronomer aiming an extraordinarily sharp telescope at that ship to peer through its windows at the clock). Length contraction resembles the predictions of Lorentz and FitzGerald of a squeezing in the direction of motion due to the pressure of ether. Yet, rather than being due to an external substance it represents the squashing of space itself along the path a high-speed object travels. Finally, mass-energy equivalence—embodied in Einstein's famous equation "*E* equals *m c* squared"—shows that mass and energy can freely be transformed from one into another. The base level mass for a stationary particle is called its "rest mass." When an object acquires energy (by speeding up, for example) it also gains "relativistic mass," making it heavier. Photons don't have rest mass, but they always have relativistic mass by means of their energy of motion. We note these effects today in high-energy particle experiments in which constituents (other than photons) become heavier and their decay rates slow down as they approach light speed.

Relativity also affects judgments of simultaneity. A farmer in Kansas, gazing up at the sky during a thunderstorm, might judge lightning strikes on two different barns, equally distant from him, separated from each other by a vast field, as happening precisely at the same time. But, from the vantage of a high-speed airplane flying above the space between the two barns, the strikes might not appear simultaneous, but rather the one it is moving toward might be seen first and thereby seem to take place slightly earlier than the other. Neither would be correct, as simultaneity is relative.

The order of causality, on the other hand, is universal—at least according to the standard interpretation of special relativity. As Einstein emphasized, causally connected events, such as a lightning blast spurring a fire, must transpire in the same progression according to all reference frames. No plane or spaceship might fly by and report that the fire caused the lightning blast. Such invariance of causal order corresponds to observers' moving at less than light speed—another reason to consider that velocity an upper limit for natural interactions.

From antiquity until the age of classical physics, scholars debated whether or not signals and materials might be sent at unlimited speeds. If not, then what might be the speed limit? As Galileo and others pointed out, light seemed to be one of the swiftest agents in the universe. We've seen how numerous measurements, from Rømer to Michelson, confirmed light's high, but finite, velocity. Yet until Einstein there wasn't a solid reason to exclude the notion that something might move faster than light does in empty space.

Special relativity offered ample support for the assertion that the vacuum speed of light is a firm upper limit for causal natural transactions. The time dilation and length contraction formulas, the definition of relativistic mass, and other measures each include a factor that is a real number for velocities less than the vacuum speed of light, zero for velocities equal to the speed of light, and an imaginary number for velocities greater than the speed of light.

A real number is one that can be represented as a notch on a number line. That category includes counting numbers, negative numbers, ratios such as one-third, and irrational numbers such as pi. In contrast, an imaginary number—a multiple of the square root of negative one—cannot be expressed on a standard number line. If plotted, it is typically awarded its own special number line at right angles to the real one. In general, observable physical quantities, such as weight or velocity, require real numbers to be expressed directly. Only indirectly might they be couched in terms of imaginary numbers—with their being squared or multiplied together, for instance, in such a way to get real values. Consequently, the faster-than-light solutions of Einstein's formulas, with their imaginary values, seemed physically unrealistic.

The situation is akin to saying that in an indefinitely long row of houses going from east to west, each has 1,000 square feet less living space than its eastern neighbor. If the first has a living space of 4,500 square feet, the second has 3,500 square feet, the third 2,500, and so forth. Practical considerations would limit that row to precisely five houses. Beyond the house with 500 square feet, you couldn't construct houses with negative square footage. Mathematically, to obtain a negative area for a perfectly square room, you'd need to assign imaginary numbers to the length and width. Obviously, with a standard measuring tape, that would be impossible. Similarly, Einstein's prohibition of velocities faster than the vacuum speed of light prevents getting imaginary values for measurable physical quantities.

Moreover, special relativity shows that the vacuum speed of light is attainable only by objects of zero rest mass. It would be impossible for a sub-light-speed object of non-zero rest mass, an electron, let's say, to accelerate to the vacuum speed of light. That's because Einstein's mass-energy equivalence formula shows that it would take an infinite amount of energy to allow a massive object to reach light speed. A spaceship attempting to catch up to light would burn up more and more energy as it accelerated, burgeoning more and more in its relativistic mass until, no matter how much fuel it had, it would run out before it reached its goal.

Not even the most powerful space cruiser could win a race to the moon against an ordinary laser.

Materials other than a vacuum are a different story. Various nonconducting or imperfectly conducting substances called dielectrics—such as glass, plastic, and water to name but a few—create obstacles in light's path, which can slow it down indefinitely. Then it is possible for other things to exceed light's speed in that material. The result of high-energy particles exceeding light's speed in a substance is a glow called Cherenkov radiation, discovered by the Soviet physicist Pavel Cherenkov. Note, however, that the expression "faster than light" usually means "faster than the vacuum speed of light," which is the sense in which we will be using it.

Phantom of the OPERA

Interestingly, as the Indian-American physicist E. C. George Sudarshan, his graduate student V. K. Deshpande, and others showed in the 1960s, special relativity does offer a hypothetical way of accommodating faster-than-light particles (in a vacuum), as long as they consistently remain faster than light and have no opportunity to slow down to light speed.[2] In a 1967 paper, the Columbia physicist Gerald Feinberg dubbed these "tachyons."[3] For tachyons, it takes more and more energy to slow down closer and closer to the speed of light. Thus, they are barred from the possibility of hitting light speed just as much as slower-than-light particles—called "tardyons" or "bradyons"—only from the opposite side of that hurdle.

One way of constructing tachyons would be to posit particles of imaginary rest mass. At first glance, that might seem completely unphysical. However, by traveling faster than light, an imaginary rest mass particle would have a relativistic mass and energy that are real numbers, and hence measurable. For a particle that could never stand still, its rest mass would be an abstraction anyway.

Special relativity also tells us that according to the perspective of some observers, faster-than-light interactions would take place backward in time. Thus, their existence could conceivably reverse the direction of cause and effect, violating the normal order of causality. In Einstein's mind, that ruled out the possibility of objects exceeding light speed. Arguing that causality must be unidirectional in time, he dismissed such supraluminal solutions as mathematical aberrations.

As Einstein wrote in a 1907 paper:

"This result means that we would have to consider as possible a transfer mechanism whereby the achieved effect would precede the cause. Even though this result, in my opinion, does not contain any contradiction from a purely logical point of view, it conflicts with the character of all our experience to such an extent that this seems sufficient to prove the impossibility of the assumption [of velocities greater than light speed]."[4]

The prospect of faster-than-light objects stirred the British-Canadian mycologist A. H. Reginald Buller to compose a whimsical limerick, which would appear anonymously in the satirical magazine *Punch*:

There was a young lady named Bright,
Whose speed was far faster than light.
She went out one day,
In a relative way,
And returned the previous night![5]

A 1970 paper by Gregory Benford, D. L. Book, and W. A. Newcomb, "The Tachyonic Antitelephone," grappled with the scenario of messages beamed into the past using tachyon signals.[6] Benford, a science fiction writer as well as a physicist, explored that idea further in his imaginative novel *Timescape*. What if a future society, he pondered, wished to warn the past of an impending disaster? Tachyon messages might be the way to do it.

Unfortunately for those hoping to offer stock market tips to their earlier selves, no such tachyons exist in the standard model of particle physics. Even if they somehow were identified outside the scope of that model, it is unclear how they could be used to convey information. Moreover, even if they could hypothetically carry signals backward in time, nature might realistically prohibit them from doing so to protect the laws of cause and effect. Temporally backward communication could engender paradoxes, such as telling one's younger self never to experiment with tachyon beams when he or she gets older. If such advice were followed, the warning signal might not be sent. Then how would the admonition arise? Because of such practical and conceptual issues, interest in tachyons has waned significantly since the 1960s and 1970s.

Still, unless experimental evidence absolutely rules out a phenomenon, theoretical physicists like to explore reasonable alternative views. A minority community of researchers maintain that tachyons remain within the realm of possibility. Just because they haven't been found yet and are not part of the standard model, those dissident thinkers argue, doesn't mean we should stop searching for them experimentally and abandon considering them theoretically.

In 1985, the physicists Alan Chodos, Avi Hauser, and Alan Kostelecky speculated that one of the "flavors" (particle types) of neutrinos (electrically neutral, lightweight elementary particles) could conceivably represent a tachyon.[7] Chodos explained his reasoning, pointing out that he meant the faster-than-light variety, not the more contemporary usage in string theory:

"If tachyons exist there are two possibilities, either tachyons are an entirely new species that remain to be discovered or one (or more) of the already known species of particle is tachyonic. If the former, discovery of tachyons would be a long way off, and their influence on current physics would be small. A more attractive possibility is that we have already 'discovered' tachyons even though we have not yet realized it. Presumably the masslessness of gauge bosons (photon, graviton) is protected by gauge invariance, so the only 'realistic' possibility is that one, or more, of the neutrino flavors is tachyonic."[8]

Because of their light weight, electrical neutrality, and complete unresponsiveness to the strong nuclear interaction, neutrinos are among the least interactive of all elementary particles. Myriad such particles pass unhindered through the Earth each second. Thus, if a renegade type of neutrino were indeed tachyonic, conceivably it might have evaded detection so far. Moreover, even if a hypothetical tachyonic neutrino breed violated the usual order of causality, it is possible that its backward-in-time influence was too subtle for particle physicists to notice. As Chodos pointed out:

"If neutrinos are the only known particles to exhibit superluminal behavior (perhaps accompanied by others yet-to-be-discovered) then any manifestations of causality violation would hardly affect our daily existence. It would be very difficult to kill your grandfather using a beam of neutrinos."[9]

In 2011, a research team called OPERA (Oscillation Project with Emulsion-tracking Apparatus) issued the startling announcement that it had measured the time of flight of neutrinos produced at the LHC (Large Hadron Collider) and detected at the Gran Sasso Laboratory, and determined that the particles traveled very slightly faster than light. The LHC is situated in a large circular tunnel that runs beneath the Swiss-French border, roughly 450 miles away from the Gran Sasso mountain in central Italy, where the OPERA detector was housed in a tunnel. During their flight path, the neutrinos took about 60 nanoseconds' (billionths of a second) less time than they should have at light speed. The team asserted that they had eliminated likely sources of experimental error before reaching their startling conclusion.

While much of the physics community was dubious, holding back judgment until independent confirmations by other teams, the story received considerable press coverage. "Faster-than-light Neutrino Poses the Ultimate Cosmic Brain Teaser for Physicists," proclaimed a *Washington Post* headline. The article went on to state that physicists have "been presented with an experimental finding that threatens to blow their vision of the universe to smithereens."[10]

In popular forums, prognostications of the imminent fall of special relativity abounded. Pundits prepared to bid farewell to Einstein's theory, which, by then, had stood for more than a century. On the lighter side, the purported discovery spurred a brief social media craze for superluminous neutrino jokes, such as: "'We don't allow faster than light neutrinos in here,' said the bartender. A neutrino walks into a bar," and "I wrote a speed of light joke . . . but a neutrino beat me to it."[11]

Several months later, the OPERA team sheepishly admitted that it had blundered. Further review of the apparatus had revealed problems with the timing system, including a loose connection. Rather than a dazzling harbinger of revolutionary physics, the "superluminal measurement" was a mere illusion born from systematical error: a phantom of the OPERA detector, so to speak.

Meanwhile, a competing team, ICARUS (Imaging Cosmic and Rare Underground Signals), whose detector was also located in the Gran Sasso tunnel, performed similar measurements of LHC-produced neutrinos, compared them to speed-of-light predictions, and found that they arrived right on time. "Had we found 60 nanoseconds, I would have sent a bottle of champagne to OPERA," Nobel laureate and team spokesperson Carlo Rubbia remarked—anticipating he'd be raising a glass to Einstein instead. "It's quite a relief, because I'm a conservative character."[12]

Despite the OPERA debacle, Chodos has remained enthusiastic about the search for tachyons in nature. He has continued to publish about the possible ramifications of their finding, include the possibility of a new symmetry in nature, called "Light Cone Reflection," that would link particles traveling above and below the speed of light.

"Most people regard tachyons as unacceptable," he remarked, "because of alleged paradoxes associated with causality violation. I do not feel that way. I think the discovery of tachyons would shake up our notions of space-time in a good way, and that nature would find a way to resolve any logical paradoxes."[13]

The Cosmic Tapestry

Space-time, the melding of space and time into a single, four-dimensional entity, offers a convenient way of discussing questions of causality and information transfer in relativity. Unlike the four-dimensional visions of nineteenth-century mathematicians, instead of a spatial fourth dimension, time serves that role. The mathematician Hermann Minkowski (who happened to be one of Einstein's former professors) proposed such a four-dimensional unification in 1907 upon finding that he could express special relativity more elegantly in terms of a unified space-time, rather than space and time considered separately. He proclaimed the union as a revolution in scientific understanding.

A space-time diagram (also known as a Minkowski diagram), depicted with time as the vertical axis and one of the directions of space as the horizontal, offers a clear picture of the limits of causality. Criss-crossing such a diagram, through any given point, representing the time and place of any given event—a New Year's Eve television broadcast from Times Square, New York, for example—is an X-shaped object, called a "light cone," that delineates all the other points the light ray might reach at other times. The reason the X-shape is called a "light cone" is because, when rotated around yet another axis, representing a second spatial dimension, it looks like an upright ice cream cone resting on top of an upside-down cone, to form something like an hourglass shape. If a point lies on the same light cone as the original event, a light signal would be able to convey information between the two of them. Such a transmission is called a "lightlike" interaction.

Within the light cone are a host of points that are closer to the original event than light would travel in a certain time. That region, called "timelike," represents all the possible slower-than-light connections. For example, for a midnight rock concert, it would include all the places sound waves, fireworks, people fleeing from the noise, and so forth, might reach within a given time frame. Both "lightlike" and "timelike" interactions are within the range of causality. Someone might see a flash, hear a sound, or be physically knocked over, and be affected by the source of the disturbance. Each of those possibilities would be represented on or within the light cone.

Finally, along with "lightlike" and "timelike," there is a third region of the space-time diagram, called "spacelike," designating all those points beyond the range of causality. For instance, an astronaut on board a spaceship near Proxima Centauri, the closest star other than the Sun, could not possibly know the set list of a New Year's Eve concert until a few years later (unless it happened to be published in advance). It would take at least that long for the signal to be broadcast that far.

Minkowski brilliantly showed how special relativistic length contraction and time dilation might be depicted using a space-time diagram. To model one observer traveling at a high velocity relative to another, simply tilt the space and time axes of the first relative to the second, like twisting an ice cream cone to the right or left in one's hand. Compared to the tilted cone, even though their space-time distances would be the same, the upright cone would be longer in the time direction and shorter in the space direction. Therefore, it shows how if time is stretched according to a certain measurer, length contracts—much like adjusting a faucet lever to increase the hot water and decrease the cold.

By rendering time similar to space, Minkowski's interpretation of Einstein's theory essentially froze the past, present, and future into a single slablike reality—called the "block universe." Thus, the past and future became theoretically (if not practically) as accessible as the present. Laplace's demon wouldn't need to calculate the future. In theory, he could simply step outside of the block universe and see all of eternity at once.

Relativity's ironclad determinism satisfied Einstein's sensibilities. Free will, he fervently believed, was an illusion. Chance elements of a theory, he argued, stemmed from lack of knowledge, rather than any fundamental randomness in nature. Therefore, he was pleased by relativity's rigid connections between the past, present, and future.

To allow for accelerating systems and model the effects of gravitation, Einstein spent an arduous decade testing various possibilities. The result was his masterful general theory of relativity, completed in 1915. General relativity showed how mass and energy warp the space-time in their vicinity, altering the paths of objects in that region. Effectively it replaced Newton's vision of gravitation acting over a distance with a local, geometric field theory.

To understand Einstein's motivation for developing the general theory of relativity, imagine that one day the Sun suddenly vanished. Astrophysics does not predict such a sudden disappearance, but let's consider it for argument's sake. Because of the time needed for light to travel through space, those on Earth would observe its blinking out roughly eight minutes after it actually happened.

Newtonian theory suggests that immediately upon the Sun's extinction, the invisible gravitational cords connecting it with each planet would be cut, and Earth and its sister worlds would each go flying into space, traveling along straight lines. The breakage would transpire instantly, not waiting for the last of the Sun's rays to arrive. Thus, it would happen faster than the speed of light, a violation of the causality limits built into special relativity. How would Earth, then, realize that it needed to change its course of motion before any signals at all could arrive from the Sun?

To rectify that paradoxical situation, Einstein realized that he would have to construct a relativistic field theory of gravitation. The simplest model of an energy field's conveying gravitation would be via ripples in the fabric of space-time itself. That formulation was possible because of the "equivalence principle," a proposal Einstein developed for equating curved motion due to gravity with the effects of an accelerated reference frame.

The equivalence principle popped into Einstein's imagination when he was thinking about someone standing on the roof of their house, losing his footing, and plunging toward the ground in a free fall. If they happened to be carrying an object—a toolbox, say—and let it go, it would drop down right alongside him as if they were both at rest. The reason the two would keep pace was Galileo's discovery that in the absence of air resistance, falling objects reach the ground at the same time, independently of their masses and other properties. Einstein took that concept a step further. Within an accelerated reference frame that kept pace, something like an encapsulating, invisible elevator, the plummeting man and the object he was carrying would seem to behave as if they were in a state of inertia. No local experiment would be able to reveal the difference between a free fall and being at rest in empty space. Inertia, therefore, could be defined locally at each point in space in terms of a freely falling framework. Einstein would later refer to that insight as the "happiest thought of my life."[14]

The equivalence principle allowed Einstein to pretend that each point in space, modeled by something like an invisible, free-falling elevator, was locally at rest. Therefore, he could apply the rules of special relativity, designed for non-accelerating systems, by setting up a space-time diagram at every point, in the manner of Minkowski, but expressed in terms of the coordinates of the freely falling framework rather than a fixed frame. The next step, he realized, would be to sew all of the local patches into a seamless garment. The needle and thread he used were the relationships of differential geometry, a higher branch of mathematics that combined calculus with non-Euclidean geometry. In non-Euclidean geometry, rules governing points, curves, and shapes—such as the definition of parallel lines and the sum of angles in a triangle—become distorted by the warping of space.

General relativity thereby behaves like a universal sewing machine that stitches patches of space-time together according to rules that vary from point to point. Those local rules depend on the mass and energy in that region. Thus, the distribution of mass and energy throughout the cosmos informs how space-time twists, turns, and curves. That space-time fabric, in turn, sets the limits of causality in each place, via the orientation of light cones, as well as the behavior of bodies passing within each region. For example, the warping of space-time by the Sun causes each of the planets to follow elliptical orbits as they ride along the crests of its gravitational well.

Einstein's general theory of relativity offered several key predictions. One was a precession (angular movement) of the orbit of Mercury. Each time Mercury revolves around the Sun, its orbit advances slightly in a manner consistent with general relativistic calculations. Einstein's estimates nicely matched astronomical readings.

Another prediction was the bending of starlight by massive objects such as the Sun. Newton's corpuscular model of light also made a prediction that starlight would bend, but by half the angle that Einstein calculated. The effect could be tested, Einstein noted, by measuring the position where a star seems to appear during a total solar eclipse and comparing that location to where it lies in the nighttime sky. The angular deviation could then be compared to the general relativistic and Newtonian prognostications to see which matches more closely.

Einstein's general theory was released in 1915, during the height of World War I. Because of the war, organizing an eclipse expedition would be daunting. A colleague of Einstein, Erwin Freundlich, tried to do so (in 1914, before general relativity was complete) and was captured and detained by Russian troops. Only after the war, in spring 1919, would a total solar eclipse in the southern hemisphere offer the opportunity for two astronomical teams to collect enough data to favor general relativistic over Newtonian predictions.

In confirming Einstein's theory over Newton's, nature demonstrated that gravitation is a local, rather than a long-distance, phenomenon. In doing so, Einstein seemed to banish remote influences once and for all. Reality consists of direct, causal chains of connections, his model emphasized. Everything else is either spurious relationships, or hidden linkages yet to be uncovered.

Cracking Open the Atom

Nature has a sly way of subverting expectations. Just around the same time that general relativity seemed to suggest that reality is an impeccably stitched garment, new models of the atom began tearing that image apart at the seams. While Democritus had coined the term "atom" to mean indivisible—something absolutely solid and impossible to cut, like a perfect diamond in the extreme—substantial experimental evidence pointed to a radically different, more fragile structure.

By the 1910s, it had already been known for some time that electrons were much lighter than atoms. Scientists also realized that neutral atoms could bleed electrons through processes such as the photoelectric effect and become positive ions. J. J. (Joseph John) Thomson, the renowned director of the Cavendish Laboratory in Cambridge, England, who had played a pioneering role in identifying the electron, had proposed a "plum pudding" model of the atom in which positive and negative charges were scattered throughout.

Though wrong about the distribution of electrons in the atom, Thomson was an excellent physicist and educator, who had a keen eye for young talent. In 1895, one of his recruits to Cavendish with the benefit of a new scholarship had been Ernest Rutherford, a homespun farmboy from rural New Zealand. When not assisting with his family farm, near the town of Nelson on the northern tip of the South Island, young Ernest had tinkered as a hobby with radio sets, cameras, and other equipment. Upon finding out about the scholarship from his mother, he had reportedly dropped the spade with which he was tilling soil and exclaimed, "That's the last potato I'll dig."[15]

Rutherford had turned out to be a prize catch. Despite being teased at posh Cambridge because of his bumpkinish background, he was a quick study and soon became a leading expert in experimental physics, particularly the study of radioactivity. After receiving his doctorate and heading to McGill University in Canada for an academic position, he returned to England in 1907 upon being appointed Chair of Physics at the University of Manchester.

The burly, energetic professor was quite a character. With an extraordinary knack for the nuances of atomic physics, he was adept at constructing just the right experiments to probe critical questions. Driven with a burning curiosity, he would sometimes get impatient with his colleagues, channeling his frustration into red-faced bursts of temper. Luckily the storms would soon pass and he'd be back to his jovial, generous self.

The biochemist Chaim Weizmann, who would become the first president of Israel, once compared Einstein and Rutherford, each of whom he had befriended at different phases of his life: "As scientists the two men were strongly contrasting types—Einstein all calculation, Rutherford all experiment. The personal contrast was not less remarkable: Einstein looks like an etherealized body, Rutherford looked like a big, healthy, boisterous New Zealander—which is exactly what he was. But there is no doubt that as an experimenter Rutherford was a genius, one of the greatest. He worked by intuition and whatever he touched turned to gold."[16]

As fortune would have it, Rutherford was lucky at Manchester to "inherit" the assistance of a skillful veteran of the lab, Hans Geiger, a German-born expert in crafting particle detectors. His clever apparatus, the Geiger counter, would become a ubiquitous means of measuring radioactive decays. Then in 1909, Ernest Marsden, a hawkeyed English undergraduate, joined the lab. Just twenty years old at the time, he had the visual acumen to spot the flashes of light from scintillators, a kind of material that records the impacts of tiny subatomic particles. The team would be a crack combination for probing the deep mysteries of atomic structure.

With a supply of radium in hand, Rutherford designed a monumental experiment that perfectly matched the skills of Geiger and Marsden. With the goal of testing Thomson's model, he crafted a system for aiming alpha particles (positively charged subatomic clusters emitted from radioactive materials, now known to be the nuclei of helium atoms) at gold foil to probe its structure. Not very confident about the outcome, Rutherford and Geiger thought that at the very least it would be a chance for Marsden to get his feet wet in the field of experimental physics.

Much to Geiger's amazement, he and Marsden found something very odd about the alpha particles' behavior. The overwhelming majority of the hurled particles they observed passed right through the gold foil as if it were baseballs flung through an open window. Yet a tiny minority bounced directly back at sharp angles, as if somehow hit by tiny batters. Stunned by the rare, but powerful, recoils, Geiger shared his excitement with Rutherford, who soon realized what was going on. Atoms, such as gold, he noted, are mainly empty space, punctuated by tiny positive nuclei. The science of nuclear physics was born!

"It was quite the most incredible event that has ever happened to me in my life," Rutherford later recounted. "It was almost as incredible as if you fired a 15-inch shell at a piece of tissue paper and it came back and hit you."[17]

Realizing that he and his team had stumbled upon an entirely new vision of the atom, vastly different from his mentor Thomson's construct, in 1911 Rutherford decided to publish his own atomic model. It introduced to the world the modern concept of the atom's having the bulk of its mass concentrated in a minute, positive central region, surrounded by mounds of empty space that somehow contained the requisite number of electrons to make the whole thing electrically neutral. It was scraped together like the radio receivers he had assembled as a kid—without much thought to the fundamental reasons it was held together. In particular, while it explained the results of the Geiger–Marsden experiment, it didn't address questions such as the stability of atoms, the nature of their spectral lines (light absorbed or released), and the circumstances in which they absorbed or released electrons, in line with the photoelectric effect described by Einstein.

Something Brilliant from the State of Denmark

Even for an insightful experimentalist such as Rutherford, sometimes it takes a theorist to resolve thorny questions. A visitor to Manchester, the Danish physicist Niels Bohr, would help resolve those issues. In spring 1912, he had arrived from Cambridge, where he had worked for about half a year with Thomson after receiving a PhD from the University of Copenhagen. Magnanimously, Thomson had told him about Rutherford's atomic model, and he was keen to meet the man himself.

In some ways, Bohr's relationship with Rutherford was like Kepler's to Tycho. Both Bohr and Kepler were quiet theoreticians, eager to interpret the data gathered by more boisterous experimentalists. In Bohr's case, luckily his mentor's results were open and relatively simple, ripe for deeper explanation.

In Manchester, and later back in Copenhagen, Bohr constructed an atomic model that in some ways resembled the solar system. Orbiting the positively charged nuclear "sun" were the negatively charged electron "planets." Rather than gravitation, what bound the system together was electromagnetism. He hypothesized that the electrons' orbits, unlike those of the planets, were perfect circles. Planets maintain orbital stability through a conservation of angular momentum—a quantity that is a product of mass, speed, and radius—along with a conservation of energy. Conservation of angular momentum is why figure skaters twirl faster when they draw their arms in closer to their bodies and slower when they extend their arms. If they do neither, they can maintain a steady pirouette. Similarly, planets balance their speeds and proximity to the Sun through angular momentum conservation to keep stable orbits. Energy conservation means they don't spontaneously get much faster and pull away, as if they were fueled rockets. Bohr speculated that each of those quantities was conserved for electrons in atoms as well.

In developing his model of the atom, Bohr came to realize that an important test would be reproducing the known atomic spectral lines of hydrogen and other simple elements. Spectroscopists such as Johann Balmer, Theodore Lyman, Friedrich Paschen, and others had measured the absorption and emission lines of hydrogen (colors of light taken in and given off, respectively, as seen in a spectroscope) and mapped out specific frequency patterns. Each spectral pattern was like a segmented rainbow, displaying only certain hues and omitting others, with frequencies that exhibited mathematical regularity. Why those colors and not others? Bohr had a hunch that electrons obeyed stationary orbits, until they absorbed or emitted photons of certain frequencies. Upon absorption or emission of a photon, the electrons would suddenly move to a higher-energy or lower-energy orbit, respectively.

In classical physics, such as Newton's description of planetary behavior, speed and orbital radius might take on any of a wide range of values. Bohr realized that to model the distinct patterns of spectral lines, electron orbits must obey a discrete, rather than continuous, spectrum. Hence, if an electron transitions from one orbit to another, that change must be a sudden jump or plunge, rather than a gradual meander.

To introduce a quantized angular momentum that would lead to particular, stable orbits, Bohr postulated that angular momentum came in multiples of a combination of constants that became known as h-bar: Planck's constant "*h*" divided by the irrational number "2 pi." Then to simulate how electrons gobbled up or released energy in the form of photons, Bohr turned to the exact formula used by Planck in his original quantum hypothesis and Einstein in the photoelectric effect: energy equals frequency times Planck's constant. Lo and behold, by combining those hypotheses with standard formulations of the strength of the electromagnetic interaction between two charges, Bohr found that he could reproduce the various spectral line formulas for hydrogen.

Bohr published his revolutionary findings in 1913. He also sent a summary of his results to Rutherford. Ever the practical one, Rutherford had a nagging question: How does an electron know when to stop when it transitions between various orbits? What is to stop it from always plunging to the lowest orbits, for example? In reality, not all conceivable shifts occur. What favors some transitions over others?

"It seems to me that you have to assume that the electron knows beforehand where it is going to stop," Rutherford wrote to Bohr.[18]

Bohr truly didn't know how to respond to Rutherford's perceptive critique. It would take a more advanced type of quantum theory—the full-blown quantum mechanics developed in the mid-1930s—to address Rutherford's comment fully.

The notion of spontaneous, instantaneous quantum leaps appeared to stand in stark contrast to the carefully plotted space-time diagrams of Einstein and Minkowski. While the former seemed haphazard, the latter set rigorous limits for causal connections. Only in the 1940s would the American physicist Richard Feynman show how such diagrams could be adapted to fit the quantum world, by inserting a measure of haziness about particle paths through space-time. Until then, Einsteinian space-time and quantum interactions practically seemed to speak in different languages.

Magic Numbers

Picture an atom, and you likely imagine intersecting ovals, oriented at several different angles, rather than a bull's-eye pattern of concentric circles. That's because the most popular image is of Arnold Sommerfeld's modification of Bohr's model, rather than of Bohr's original 1913 proposal. Sommerfeld was a gifted physicist, working in Munich, Germany, who found and rectified several major gaps in Bohr's construct.

In particular, the situation Sommerfeld addressed involved atoms being placed in the path of significant magnetic fields (such as electromagnets created by passing currents through coils of wire). According to the Zeeman effect, named after the Dutch physicist Pieter Zeeman who discovered it in 1897, turning on such fields engenders a splitting of spectral lines. Where there might have been a single stripe of a certain color, there are multiple lines of slightly different hues. It shows that a magnet acts, in a sense, like a prism, breaking up a uniform band of light into many. Why a magnet would turn one tone into a miniature rainbow was a true mystery—until Sommerfeld found the answer.

Study of the Zeeman effect would prove critical for the generalization of Bohr's simple "solar system" model of electrons circling the nucleus into a much richer three-dimensional atomic description that includes multiple quantum numbers and other features. (Investigations of the anomalous Zeeman effect, a related phenomenon that applies to atoms with odd numbers of electrons, would help complete the picture.)

The revised atomic model would allow for more precise predictions, ultimately revealing the full gamut of quantum phenomena including acausal features such as entanglement. Sommerfeld's work would thereby offer a vital bridge between Bohr's rudimentary model and the full theory of quantum mechanics, with all its weirdness, soon to come.

Sommerfeld proposed that certain orbits—corresponding to energy levels—in the original Bohr model weren't single quantum states at all, but rather were a cadre of degenerate quantum states that happened to reside together under special circumstances. "Degenerate" in this context doesn't mean "riffraff," but, rather, distinct quantum states that happen to share the same energy. Given that energy and frequency are associated by Planck's relationship, that explains why transitions to those states might be characterized by a single spectral line.

Even if a group of quantum states all have the same energy, Sommerfeld postulated, they might not possess exactly the same angular momentum. For orbits, the amount of angular momentum affects how stretched out they might be—from purely circular to highly elliptical. Therefore, Sommerfeld argued, certainly energy levels might be associated with a range of orbital shapes, rather than just a simple circle. Moreover, those orbits might be tilted at different angles relative to a central axis—generally labeled the "*z* axis" for consistency. In that case, even for two states with the same total angular momentum (overall shape), the *z* component of the angular momentum (orientation relative to the *z* axis) could still be different. In short, rather than being distinguished by just energy levels, as Bohr surmised, Sommerfeld added two additional parameters: total angular momentum and *z* component of angular momentum.

To characterize the full range of possibilities for electron states, Sommerfeld assigned them three different quantum numbers: principal (the original Bohr levels, connected with energy), azimuthal (related to the total angular momentum), and magnetic (related to the *z* component of angular momentum). These are the "street names," "building numbers," and "apartment numbers" for electrons, setting their addresses within the atom. For degenerate states, multiple sets of quantum numbers share the same overall energy. However, an external influence, such as a magnetic field, might "couple" (connect as a force) with the specific angular momentum states, splitting their energy levels and yielding a range of spectral lines where otherwise there might have been a single one.

It is like a supermarket in which customers have, at first, absolutely no preference for apple varieties. Therefore, all the apples, no matter whether they are McIntosh, Golden, Gala, Delicious, or Granny Smith, are sold from the same bin at exactly the same price: twenty-five cents. On busy days, customers form a single line for the solitary apple bin, pick one up at random, and pay a quarter for each one at the counter.

Suppose, however, a new study comes out that says Golden apples have extraordinary nutritional properties. Meanwhile, reports emerge that one of the other types soaks in pesticides and is not so healthy. The store manager might decide—based on external factors—to split the apples into multiple bins and charge different prices for each variety, leading on busy days to multiple lines. Similarly, the magnetic field's influence serves to distinguish various electron states, turning a single energy "bin" into multiple bins.

In 1916, Sommerfeld concocted a new natural constant to characterize the coupling between electrons (and other charged particles) and photons, as they interact to yield electromagnetism. He defined it as the speed of an electron in its lowest-energy, relativistic atomic state, divided by the speed of light. Called "the fine structure constant" or "Sommerfeld's constant," it is a dimensionless combination of electric charge, Planck's constant, and the speed of light. "Dimensionless" in this context means a unitless number—unlike, for example, time, which has units of seconds, mass, which has units of kilograms, and so forth. Therefore, for any system of units, the fine structure constant always has the same value. Strangely enough, and seemingly by pure coincidence, that value is very close to (but not exactly) the number 1/137. Why the reciprocal of the particular whole number 137, which happens to be a Pythagorean prime number (a sum of two squares, divisible only by the number 1 and itself) among other curious mathematical characteristics? Surely combining three fundamental constants—electric charge, Planck's constant, and the speed of light—would not yield something so simple. Starting with Arthur Eddington (who erroneously thought that the fine structure constant was exactly 1/137), many physicists would ask that question, looking for connections with other aspects of nature, and wrack their brains searching for a meaningful answer. Numerical patterns can, in some cases, lead to spectacular insights. For instance, the electron quantum numbers relate closely to chemical properties, as indicated by the periodic table of elements. Other times, though, too much rumination about certain numbers offers a numerological dead end. We've seen how Kepler sought unsuccessfully to map the orbits of planets onto embedded Platonic solids. Contemplation of the significance of the number "137" has offered another notable quixotic adventure in the annals of science. It would become an irresistible challenge—for some, even an obsession.

In modern physics, the balancing act between pure empiricism and mathematical abstraction has proven tricky, no doubt. Sway too much toward demanding that everything is objectively measurable—that is, toward realism—and one misses nuances such as quantum leaps. Veer too much toward mathematical elegance—that is, toward idealism—and one loses touch with credible experimental verification. Somewhere in between is the optimal path toward continued progress in physics.

5

THE VEIL OF UNCERTAINTY

Turning Away from Realism

> *To me a deterministic world is quite abhorrent—this is a primary feeling. Maybe you are right, and it is as you say. But at the moment it does not really look like it in physics—and even less so in the rest of the world. I also find your expression, the 'dice-playing God', completely inadequate. You have to throw dice as well in your deterministic world; this is not the difference . . . I think . . . that you underestimate the empirical fundamentals of the quantum theory.*[1]
>
> —MAX BORN (letter to Albert Einstein, 1944)

SEEING IS BELIEVING. OR IN CONTEMPORARY TERMS, PICS OR IT never happened. We trust our sensory perceptions to gauge and affirm the conditions of the natural world. We know from childhood that pushes and pulls have something to do with motion. Although our hunches about what causes things to move might misguide us into taking the Aristotelian view that forces relate directly to speed, it is relatively simple to accept the Newtonian conclusion that forces engender *changes* in motion, that is, acceleration. Perhaps it isn't much of a stretch beyond that to adopt the system of Maxwell and others in which forces are conveyed by fields, such as electromagnetism, that propagate through space. After all, although we might not see the wind, we might hear its shriek and notice when it rips branches from trees and blows leaves through the air. Similarly, we might envision how unseen waves conveying force might ripple from one place to another.

In contrast to the realm of natural perception, there is the world of dreams. Our somnolent theatrics might include visits from long-lost relatives, unlikely encounters with celebrities, and extraordinary abilities, such as the power to soar through the air like Superman. We might leap instantly from one place to another or suddenly find ourselves in the past or a projected future. If we consider ourselves savants, we might assert that we can prognosticate the state of things to come or communicate with someone using only our thoughts. However, in waking life—put to the test of scientific rigor—any such claims would not hold up—except perhaps through spurious coincidences that could not be reproduced. So-called psychics who make numerous forecasts are bound to get some of them right through pure chance, for instance.

When Albert Einstein became world-famous, one of his challenges was distinguishing relativity—a scientific theory with testable results that happened to be couched in four-dimensional mathematics—from public misconceptions about supposed connections between higher dimensions and the occult. He stubbornly emphasized the realism of his theory—which offered exact physical predictions for any point in the universe given the state of matter and energy in that region.

Yet, much to Einstein's chagrin, quantum mechanics would veer in a wholly different direction. It would allow for random, sudden leaps and long-range connections that had no direct, causal linkage. It would replace the idea of objective physical parameters with quantum states that yield information only upon measurement. Moreover, the uncertainty principle would guarantee that certain pairs of factors, such as the position and velocity of a subatomic particle, could not be known precisely at once. In short, it would shred the fabric of direct, objective causal connections, carefully knitted by classical physics.

That is not to dismiss the revolutionary nature of relativity. Although it maintained the rigid determinism of classical physics, it did so in a manner that shattered the traditional distinction between space and time. Much to Einstein's bafflement, it would excite the public in an unprecedented way—from the strictly scientifically minded to those who embraced pseudoscientific beliefs such as telepathy and clairvoyance.

Albert in Wonderland

The first mention of Einstein's theory of relativity in the *New York Times* was relatively low-key—in a 1913 report about a speech by the English physicist Sir Oliver Lodge. Mentioning that Lodge was an occultist, the article emphasized his belief that even though "ether has failed to respond to the subtlest efforts made to detect its existence" it might still exist on the basis of "philosophical grounds."[2] Therefore, he opposed attempts by Einstein to impose relativity and banish the ether. The article made a valid point that ether (once thought to be an actual, lightweight substance) had become associated with occultism, whereas the pure emptiness of the spatial vacuum (once thought to be impossible for conveying waves) had become credible science.

Six years later, the confirmation of one of general relativity's key predictions, the bending of starlight by massive objects such as the Sun, bolstered Einstein's fame considerably, rendering him an international scientific superstar. On May 29, 1919, two British expeditions, organized by Eddington and Frank Dyson under the auspices of the Royal Society, observed a total solar eclipse from different locales in the southern hemisphere—Sobral, Brazil, and Príncipe, an island off the coast of West Africa—and noted slight changes in the positions of the stars in the region of the sky near the occluded Sun. After analyzing the data, the Royal Society declared at its November 6 meeting that they matched Einstein's predictions more closely than a Newtonian calculation (based on his corpuscle theory) would project. In short order, Einstein's victory was splashed all over the headlines. "Revolution in Science. New Theory of the Universe: Newtonian Ideas Overthrown,"[3] announced the *Times* of London. "Lights All Askew in the Heavens . . . Stars Not Near Where They Seemed or Were Calculated to Be, but Nobody Need Worry,"[4] proclaimed one of many headlines about the subject in the *New York Times*.

The floodgates were open for public exposure to the new science, including encounters with the pliable aspects of reality, as incorporated into the notion of four-dimensional space-time. Readers soon learned that space, time, mass, energy, exactitude, chance, and the fundaments of reality were not what they had seemed to be.

The firewall that science had carefully constructed in the late nineteenth century between the tangible and the mystical—distinguishing the measurable, predictable realm from speculations about unseen dimensions, ghostly movements through seemingly impenetrable barriers, and so forth—no longer seemed so solid. While new characterizations of the bounds between genuine science and pseudoscientific musings would eventually emerge, in the interim many pundits wondered if, like some of the strange fashions of that era, "anything goes."

At the very least, educated readers needed to grapple with the daunting notion that Einsteinian relativity robbed space and time of their independent definitions, replacing them with a four-dimensional amalgam. They also needed to wrap their minds around the idea that as a consequence, mass and energy were different manifestations of the same thing, as expressed in Einstein's famous equation. Like a genie, the German genius had seemingly transformed science into a bazaar of the bizarre.

A *New York Times* article published in 1923 compared the landscape of relativity to the strange, dreamlike terrain of *Alice in Wonderland*:

"Unconsciously, perhaps, through the medium of a child's dream, [Lewis Carroll] set forth a mathematician's longing for a frame of space and time not limited to three dimensions. Clearly the vagaries of Alice would cease to be vagaries in a four-dimensional existence . . . Topsyturvydom may be a land not too far distant from Relativity."[5]

While developments in relativity could be confusing, at least its dynamics obeyed mechanistic rules. Nothing about it was haphazard. Readers who didn't fathom Einstein could at least take comfort in the fact that knowledgeable physicists and mathematicians could reliably make predictions based on his work, such as the solar eclipse measurements of 1919 and a further confirmation during the solar eclipse of 1922.

Einstein's revolutionary findings transformed him in the public mind to something like a prophet. Yet there was no mysticism involved. Einsteinian relativity followed a precise, logical scheme. Rather than making physics more mystical, in many ways it made physics more solid. Not only did it abolish the vague notion of ether, it also rendered the fourth dimension more immediate by associating it with time. Moreover, it eliminated the ambiguities connected with Newtonian mechanics, such as the need for "absolute space" and "absolute time"—fixed rulers and clocks that Newton required to define inertia but could never fully explain. Therefore, if readers took time to think Einsteinian relativity through, they'd realize that it actually placed physics on firmer ground.

Quantum mechanics seemed in many ways even more mysterious than relativity. Many articles from that era emphasized how nebulous and indeterminate it was. Rather than oiling the machinery of physics, making it run smoother, it seemed to throw a cog into the works. By 1931, after years of baffling pronouncements issued by the quantum physics community, in a piece titled "How to Explain the Universe: Science in a Quandary," the *New York Times* science columnist Waldemar Kaempffert decried:

> That mechanical universe is now gone. Science is forced to become idealistic . . . The laws governing these quantum phenomena can be written down, but they are unintelligible at present . . . The truth is that the nearer we come to reality, to the rock bottom of the universe . . . the more baffled we are . . .
>
> We have adopted the scientific method because it was the easiest. It has taken about a million years to develop the scientific method to its present perfection. Suppose we were to spend the next million years in developing other methods vaguely described as intuitions, "inner voices," and the like. Would we see deeper? . . . Possibly poets and seers who have had religious experiences may be aware in a dim way of what lies behind the mathematical concepts in which we express what little we know of the world around us.[6]

Which mysterious aspects of quantum mechanics prompted such pessimism about the scientific method? For one thing, the uncertainty principle, advanced by the German physicist Werner Heisenberg in 1927 on the basis of mathematical relationships he developed earlier, demonstrated that on the subatomic level, certain paired properties, such as position and momentum (mass times velocity), and time and energy, could not be measured precisely at the same time. That is, the more one tries to pin down the position or duration of a particle, the less one knows about its velocity or energy. No more could physicists aspire like Laplace to predict all aspects of every particle in the universe. On the most fundamental level of nature, there would always be unknowns.

Complementarity, a principle advocated by Bohr, proposed a different kind of duality: between the particle and wave aspects of objects. Sometimes photons, electrons, and other subatomic constituents act like particles. For example, they scatter off of other particles at predictable angles. Other times the same constituents act like waves. For instance, the peaks and valleys of their oscillations either add up or cancel each other in the process of interference, depending on their alignment, creating striped patterns of light (heightened) and dark (weakened) signals. Bohr argued that the apparatus chosen by the observer determined whether the particle or wave properties were expressed. Until a measurement was made, the system was like a black box, revealing nothing of its contents. The experimenter's choice would then reveal certain types of results—either clumpy like a particle or spread out like a wave. No one, however, could probe the system to reveal every scrap of information at once—the "black box" remained forever closed.

Another *New York Times* article, published in 1933, tried to explain this strange situation to readers by drawing an analogy with Robert Louis Stevenson's classic tale "The Strange Case of Dr. Jekyll and Mr. Hyde":

> Professor Bohr, after a lifetime of contemplation of both the ponderables and the imponderables of the physical and mental world, has come to discover an inherent essential duality in the nature of things, as they relate to man's ability to know them. The paradox of this duality lies in the fact that the Jekyll-Hyde nature of all things is essentially contradictory, with both aspects being true at different times, but with only one aspect being true at any one given time.[7]

The article also mentioned that "Bohr told the story of how he and Einstein had recently joined intellects to find a way somehow to get around the troublesome principle of uncertainty." That claim of cooperation was likely an exaggeration. Rather, from the late 1920s onward, Einstein had broken almost completely with the quantum community, including Bohr and Heisenberg, due to its abandonment of the notion of objective physical reality. Even if our measuring devices are flawed, Einstein believed, the position, velocity, and other physical properties of objects at any point in time must have independent values. Neither the uncertainty principle nor the complementarity principle acknowledged an objective reality separate from how we measure it.

Einstein was also troubled by the concept of discrete jumps from one quantum state (representation of a set of physical properties) to another, believing that continuous equations must underlie such behavior. Heisenberg's formulation of quantum mechanics, focused initially on the behavior of electrons in atoms, calculated the probabilities that an electron would transition from one rung to another in the ladder of energy levels of an atom. Such shifts would happen instantly and randomly, with no intermediate steps—like the jerky movements from one frame to the next in early films. Einstein hoped for a smoother, deterministic underlying explanation of why such sudden transitions transpire—like the fundamental dynamic rules governing a roulette wheel that might predict on which number a ball might seemingly "randomly" land. Ultimately, he refused to abandon the mechanistic universe, with causal connections and the speed-of-light limit.

Despite his reasoned critique, Einstein never asserted that quantum mechanics was wrong. He lauded its experimental triumphs. Rather, he argued that quantum mechanics was *incomplete*. A deterministic theory of nature was needed, he believed—likely an extension of general relativity that encompassed electromagnetism and all atomic phenomena, as well as gravitation—to explain the deeper mechanisms underlying the illusion of quantum fuzziness and randomness.

As Einstein wrote to the quantum physicist Max Born in 1926: "Quantum mechanics is certainly imposing. But an inner voice tells me that it is not yet the real thing. The theory says a lot, but does not really bring us any closer to the secret of the 'old one'. I, at any rate, am convinced that He is not playing at dice."[8]

Quantum Quandaries

One of the bridges between the "old quantum theory" of Planck and Bohr—for which Einstein played a major role (in advancing the photoelectric effect, statistical treatments of quantum systems, and other ideas)—and modern quantum mechanics was Sommerfeld's training of a new generation of highly inquisitive young physicists. Among those prodigies were Heisenberg and Wolfgang Pauli, a *wunderkind* from Vienna, who happened to be the godson of the philosopher Ernst Mach. In appearance, the athletic, boyishly handsome, clean-cut Heisenberg could not be more dissimilar to the bug-eyed and doughy-framed Pauli.

Pauli made his mark at the age of twenty writing an excellent review article about relativity. It appeared in a widely noted mathematical encyclopedia edited by Sommerfeld. With insight well ahead of his years, Pauli commanded great respect. Because he was often right in his theoretical hunches, he brimmed with confidence—even arrogance—about his assertions. The theoretical community often turned to him whenever there was a tough question needing a definitive answer. For his brilliance, he would come to earn the nickname "*Zweistein*" (Einstein II).[9]

Though only about a year older than Heisenberg, Pauli often acted like a mentor, doling out suggestions—such as advising him to avoid relativity, as there wouldn't be many good projects in that field compared to atomic physics, where there seemed to be more avenues to explore. Given Heisenberg's future accomplishments in atomic physics, Pauli turned out to be most perceptive. Throughout their careers, Heisenberg would invariably check with him before making decisions related to physics, such as submitting an article for publication.

Over time, Pauli got into the steady practice of offering blunt, unsolicited advice to the countless theoreticians that happened to cross his path—during a conference, seminar, or other venue. Young researchers who attempted to explain their ideas to him would come to expect a blustering critique. Though often useful, it still stung. Thus, he would become legendary for his barbs, as much as for his brilliant contributions.

Sommerfeld was proud to introduce his clever young scholars to the older generation. His promotion of such intergenerational interplay would prove crucial to the further development of quantum physics.

In a 1922 letter to Sommerfeld, Einstein praised his mentorship of young minds:

"What I particularly admire about you is how you raised such a large number of young talents out of the bare earth. This is something unique. You must have the gift of ennobling and activating the minds of your auditors."[10]

In turn, Sommerfeld often spoke highly to his students—including Heisenberg—about the extraordinary genius of Einstein and Bohr. Brash, young, and ambiguous, however, Heisenberg respected those "elders," but wasn't inclined to have them instruct him on how he should think, whom he should follow, and what the laws of nature were all about.

While respecting Einstein's pivotal contributions toward relativity, Heisenberg would come to deride him for his philosophical blinders. Heisenberg couldn't understand why Einstein failed to recognize quantum truths. Heisenberg's impatience was understandable. Science eschews prejudices. If the fundamental rules of how things work were jumpy, fuzzy, and abstract, Heisenberg argued, so be it. Humanity should not tell nature what to do.

Heisenberg was born on December 5, 1901, in Würzburg, Germany, to Annie Wecklein and Dr. August Heisenberg, a Greek scholar who would later be appointed Professor of Middle and Modern Greek Languages at the University of Munich. For that reason, and also because of *Gymnasium* (high school) requirements, young Heisenberg was familiar with the classics, including some of the writings of Plato.

According to the physicist and philosopher C. F. von Weizsäcker, as recounted by the Heisenberg biographer David C. Cassidy, "Heisenberg probably read only the school-assigned readings, such as the *Symposium*, *Apology*, and parts of the *Timaeus*. In his view, what interested Heisenberg most was not the philosophy but the beauty and poetry of the Greek prose (students had to learn Greek at the Gymnasium)."[11]

German quantum physicist Werner Heisenberg (1901–1976), who developed the uncertainty principle; Credit: Photographer Friedrich Hund, AIP Emilio Segrè Visual Archives, with permission of Jochen Heisenberg.

But Heisenberg didn't just sit in his house and read. On the contrary, he was an athletic lad, who relished taking long hikes along mountain paths. He joined a scouting group called the Pathfinders, which further infused him with a love for nature's wild beauty.

In 1919, when Munich was faced with an attempted Communist uprising, Heisenberg was called to serve. One summer night when on duty, he revisited one of Plato's classic works and noted its references to atomism. He mentioned those ideas to his friends. As he recalled:

"[S]omehow I couldn't sleep, so I went up to the roof of the house into the sunshine; it was nice and warm. I had Plato's Timaeus with me. I studied the Timaeus of Plato partly to keep up with the Greek because I had to know Greek for my examination, but partly also because I was really fascinated by atomic theory. You know that all the atomic theory of Plato was in the Timaeus."[12]

In his later years, Heisenberg cited the *Timaeus* of Plato—which also alluded to the works of Pythagoras as well as atomism—as one of his influences in pursuing atomic theory. Just as Plato saw the tangible, measurable world as an illusion—a mere echo of a deeper, more perfect reality called the realm of "forms"—atomic physics views quantum states as fundamental. Just as Pythagoras viewed numbers as more fundamental than material things, quantum physics is encoded in numerical parameters—such as the principal, azimuthal, and magnetic quantum numbers that characterize a state. As Heisenberg would point out:

"[T]he resemblance of the modern views to those of Plato and the Pythagoreans can be carried somewhat further. The elementary particles in Plato's *Timaeus* are finally not substance but mathematical forms. 'All things are numbers' is a sentence attributed to Pythagoras. The only mathematical forms available at that time were such geometric forms as the regular solids or the triangles which form their surface. In modern quantum theory there can be no doubt that the elementary particles will finally also be mathematical forms but of a much more complicated nature."[13]

Cassidy cautioned, however, that Heisenberg may have overstated, in his late-life recollections, the role of Greek philosophy in his early thinking:

> As far as I know, there was very little, if any, impact of Plato and other ancient Greek philosophers on his pre-World War II physics, and only a slight impact on his post-war physics. For instance, in his 1969 memoir *Physics and Beyond*, he recalled reading about Plato's notions of atoms in *Timaeus* while on military duty. [But] he also recalled rejecting Plato's naive notions of atoms at the time as "wild speculation"—hardly an encouragement to look to Plato in developing his theories.
>
> In addition, Heisenberg, an extremely ambitious young theorist, would not be guided by any philosophical notions but by what worked. In his attempts to replace the failing classical physics with quantum mechanics in the 1920s, Heisenberg's motto was "Success sanctifies the means."—regardless of philosophy or physics tradition. When he did take what we might consider a philosophical position, such as positivism in his formulation of matrix mechanics or operationalism in the uncertainty relations, it was always borne of the moment—to critically analyze failing notions and to break with them. It is unlikely that he even realized he was drawing upon a philosophical position.[14]

Enrolling at the University of Munich and conducting research under the supervision of Sommerfeld, one of Heisenberg's first projects, at just the age of twenty, was trying to understand the so-called anomalous Zeeman effect. For atoms with an odd number of electrons (such as hydrogen), it states, a magnetic field divides a single spectral line (representing a medley of degenerate states) into a rainbow of lines representing an even number of angular momentum states of different energies. The normal Zeeman effect, in contrast, pertains to atoms with an even number of electrons. In that case, a magnetic field splits a single spectral line into an odd number of lines. The problem Heisenberg tried to address was that Sommerfeld's quantum number representation, with its principal, azimuthal, and magnetic numbers, all integers, predicted an odd number of spectral lines in all cases. Why, then, did hydrogen, for example, not follow that rule? Heisenberg correctly surmised that there could be half-integer quantum numbers, as well as integers. It would take a few more years, however, for him to develop his own working model of the atom.

Sommerfeld maintained a close connection with Max Born, who worked at the University of Göttingen, well to the north in central Germany. For centuries Göttingen, a charming university town, had maintained an exalted tradition of scholarship in mathematics and sciences. There, Hermann Minkowski had unified space and time into space-time and David Hilbert had explored the axioms of mathematics and the nuances of general relativity. Born was similarly innovative and open-minded. Both Pauli and Heisenberg each spent time doing research under his mentorship.

In June 1922, Hilbert, Born, and others organized a symposium at the University of Göttingen held in honor of the tenth anniversary of Niels Bohr's groundbreaking solar system model of the atom. Known as the Bohrfestspiele, or "Bohr Festival," it featured a series of talks by Bohr himself. Heisenberg was in attendance, persuaded by Sommerfeld that it would be an excellent opportunity to get to know Bohr. Pauli attended the lectures as well. Similarly, it would be his first encounter with Bohr.

Bohr's inscrutable style of lecturing was not for everyone. As he never learned how to project his voice, audience members would practically need to crane their necks and perk their ears just to try to hear him. Moreover, although he was a brilliant thinker, the way he addressed issues was often enigmatic—leaving more questions than answers. Heisenberg, in a blaze of thought, grew impatient. When the audience was offered the chance to ask Bohr questions, his hand shot up—despite the tradition in those days for students to remain quiet in the presence of more venerable thinkers.

Heisenberg first asked Bohr about the meaning of frequencies in his atomic model. In classical physics, orbital frequency pertains to the number of revolutions per second. Why did Bohr's model ignore that definition and choose a definition for frequency that seemed to have nothing to do with orbital rate? Secondly, Heisenberg wondered if Bohr had made progress with other elements besides hydrogen (and singly ionized versions of other atoms) that would take into account the interactions between multiple electrons.

Bohr was not flummoxed by Heisenberg's probing questions—which in some ways were akin to Rutherford's critique about his model's limitations. Given that they lacked easy answers, he thought it best, however, to address them in a more relaxed manner. So, he offered Heisenberg a friendly invitation to take a walk together in the nearby hills. Heisenberg was delighted, and the two had a pleasant discussion in which Bohr admitted that there was far more work to be done in resolving the loose ends of atomic theory.

Given that, as Bohr would readily concede, atomic frequencies had nothing to do with how quickly electrons orbited, Heisenberg decided that the Bohr–Sommerfeld (Bohr's notion, as modified by Sommerfeld) visualization of electrons as akin to planets' speeding around a nuclear "sun" was not particularly useful. Rather, he set out to build a new model of electron behavior from scratch—basing it on observed atomic spectra.

Reality and the Matrix

Heisenberg's road to a new mathematical model of the atom was rather circuitous. Bohr invited him up to Copenhagen for a research visit. Sommerfeld advised him to work with Born instead at the University of Göttingen, complete his PhD (including a thesis defense back in Munich), and then only afterward consider Bohr's invitation. That's precisely what Heisenberg did. His main glitch was defending his dissertation before a committee. Although he flubbed some of the questions, due perhaps to his narrow focus, Sommerfeld still barely permitted him to pass. Once surmounting that hurdle, he returned to Göttingen, traveled up to Copenhagen for some months, and went back to Göttingen yet again to continue his research under the guidance of Born.

As he pondered atomic physics, one issue Heisenberg thought was critical was explaining why different spectral lines had different intensities, or levels of brightness. In contrast to the evenness of rainbows, in the spectrum of each chemical element, some hues shone more vividly, and others more faintly. The Bohr–Sommerfeld model, while explaining the frequencies of many of the spectral lines, did not touch on that issue.

At the time, Heisenberg was rudderless and confused. In a 1924 letter to Bohr, Pauli complained about Heisenberg's lack of direction: "He is very unphilosophical, for he does not pay attention to the clear working out of the basic assumptions and their connection with prevailing theories."[15]

About a year later, Heisenberg wrote to Pauli: "Unfortunately my own private philosophy is far and away not so clear, but rather a mishmash of all possible moral and aesthetic calculation rules through which I myself often cannot find my way."[16]

In June 1925, after churning through potential explanations of the intensity distribution over and over again until his brain ached, Heisenberg faced another kind of head pain: a severe allergy attack due to airborne pollen. Deciding to take a breather, he went on holiday to the barren, windswept northwest German island of Heligoland, nestled in the North Sea, when the cool, salty air was most invigorating. Once his hay fever subsided, the stark beauty of that red sandstone enclave lapped by the wild sea stimulated his thinking, and he developed a brilliant solution to the intensity problem.

One of the go-to analogies for theoretical physicists when faced with a dauntingly unfamiliar situation is to express its dynamics in terms of coupled (linked) harmonic oscillators—in other words, a network of springs, like an exposed box spring under a mattress. Springs are incredibly versatile in modeling a wide range of periodic phenomena: from the pendulum on grandfather clocks to rocking chairs.

Making calculations refreshingly simple, a small set of parameters describes the complete behavior of a frictionless spring. One measure is frequency—which in the case of a spring is the rate of moving back and forth during each cycle. Another is phase angle, defined as what part of a complete cycle the spring's motion begins at—fully stretched, fully compressed, absolutely lax, or somewhere in between. Yet another parameter describing springs is amplitude, the maximum amount of stretching from the equilibrium position. Squaring that amplitude and multiplying by a constant yields the spring's total energy: the equivalent of brightness for light and loudness for sound. In fact, for all three cases—springs, light, and sound—one might define intensity to be proportional to amplitude squared. Thus, amplitude squared represents a critical measure of strength.

Take, for example, a classic pinball machine with a spring mechanism that causes a ball to rise. Assume that friction has little impact. Pull the handle at the base of the machine, and a spring stretches. Release it and the spring snaps back, transforming its energy into the kinetic energy of a ball, causing it to rise. Now pull the lever again, this time twice as far from its resting position, and the spring amplitude would double. Consequently, the intensity (in this case, the total energy) would be four times the original amount, and the ball would rise four times higher than for the original pull.

While in Heligoland, Heisenberg pondered a proposal by Bohr's main assistant, the Dutch physicist Hendrik "Hans" Kramers—along with Bohr and a visiting American researcher, John Slater—to model electrons as simple harmonic oscillators. Extending that idea, Heisenberg developed schemes for multiplying amplitudes together to form probabilities in an abstract space, rather than directly measurable intensities. His goal was calculating a kind of table of probabilities for all possible transitions between quantum states. Brighter spectral lines would then match up with the net effect of more probable transitions, and dimmer lines with less probable transitions.

Just as springs can be described either in terms of their position or momentum (mass times velocity), Heisenberg explored both kinds of representations for electron states. But then he noted something curious. To get his calculations to work, he found that he needed to abandon the commutative property of conventional multiplication, which states that x times y is the same as y times x. Rather, for the notation he used, order of multiplication mattered—a situation known as "noncommutative." Specifically, multiplying a "momentum operator" (mathematical function representing the momentum) times a position state yielded a different result than multiplying a "position operator" (mathematical function representing the position) times a momentum state.

Imagine going to a strange country in which the order of multiplication matters for its currency. Suppose you go to a bazaar and toss ten coins, each labeled "two crowns," on a merchant's table and you ask him, "What can I buy with these?" He responds, "Sorry, that isn't worth much, you might buy with it only a thimbleful of rice." Then you pick up the first set of coins, pocket them, and set down two "ten crown" pieces instead. Suddenly, the merchant's expression changes and he responds, "Ah, for that amount you can have my finest handwoven carpet." What has changed? Only the order of multiplication. But as Heisenberg found, that can make a pivotal difference.

Excited by his breakthrough, Heisenberg shared his insight with Pauli, then working at the University of Hamburg, who, despite his tough standards, heartily approved. Then Heisenberg traveled back down to Göttingen, where he huddled with Born, and a young, savvy German graduate student, Pascual Jordan, to firm up the mathematics. Born considered Heisenberg's finding a key component of what he called "quantum mechanics": a system for manufacturing physically observable measures by applying operators to quantum states occupying an abstract space of possibilities.

To complete the theory, Born suggested expressing it in the language of the mathematical field of linear algebra. Linear algebra characterizes different kinds of mathematical entities and assigns rules to each for addition, subtraction, multiplication, and other operations. The simplest entity is a scalar, or ordinary number. These can be treated by means of the basic arithmetic learned in school. Physically, quantities such as the temperature of a room or the age of a fossil represent scalars. In nonrelativistic physics, energy, time, and mass are also scalars.

Constituting the next level of complexity in linear algebra are the categories known as "column vectors" and "row vectors." Column vectors, dubbed "kets" by the British physicist Paul Dirac, resemble standing dominos, or a single column in a spreadsheet, in that they offer a vertical list of numbers, labeled according to their row. Row vectors, dubbed "bras" by Dirac (put "bra" and "ket" together with a constant "c" in between to get "bracket"), offer the same, but only horizontally.

Finally, matrices offer an even more intricate mathematical entity. A matrix—something like an accountant's logbook—contains both rows and columns of relevant information about a system. Other matrices can transform it—for example, to represent physically rotating the system—via a mathematical process called matrix multiplication. Square matrices are in some ways easiest to work with, as they have symmetry in their number of rows and columns. The mode of expressing quantum mechanics using matrices is called "matrix mechanics."

As Born pointed out to Heisenberg, matrix multiplication is noncommutative; that is, order of operation matters. Therefore, matrices proved ideal for representing quantum operators, such as position, momentum, and energy (called the Hamiltonian operator). The quantum states themselves, such as the state of an electron (or pair of electrons, as the case may be) in a specific configuration within an atom, might be expressed using column and row vectors. Finally, the physical observables, such as the measurable energy of an electron's transition between different atomic levels that produces a spectral line, might be represented using scalars.

In Newtonian physics and relativity, physical parameters might be predicted exactly. For quantum mechanics, in contrast, one might only calculate the most likely value of that parameter, called its "expectation value." An expectation value is a kind of weighed average of the possible values of a certain quantity, each multiplied by its likelihood. Those probabilities, in turn, connect with a kind of "square" of the vector state, which emerges by multiplying a column vector and its corresponding row vector together. Typically, the operator connected with the physical property being measured—such as position, momentum, or energy—is set in between the column and row vectors during the weighted averaging process. In short, to obtain the expectation value of a given operator as applied to a particular quantum system, sandwich the matrix representing that operator between column and row vectors representing the array of possible quantum states in the system, and perform a weighted average. That method is certainly not as direct as simply writing down the position, momentum, and so forth, as basic variables. Yet nature, at its subatomic level, behaves according to quantum rules, not classical.

Behind the Scenes

In Newtonian physics, all of the action takes place out in the open. Kick a football and someone might record its position and speed at any given moment during its flight through the air. By weighing it and finding its mass, its momentum might be determined as well. Moreover, all of those physical parameters might be predicted as well, from Newton's laws, in a manner that is readily visualizable. Each step might be drawn in a notebook, if one is so inclined.

Not so for quantum mechanics. Its mechanisms, though well understood, are furtive and behind the scenes. Its vectors and matrices—representing quantum states and operators, respectively—live in an abstract domain known as Hilbert space, named after David Hilbert who developed the idea. Hilbert space codifies all the information that might possibly be known about quantum systems, but in a manner accessible only when targeted experiments are conducted. When a measurement is taken, mathematically, that observation is equivalent to applying the pertinent operator to the complete set of state vectors that spans all of the possible outcomes for the quantum system under study.

In essence, Hilbert space resembles a kind of network of utility corridors behind the scenes of an active office building, or the hidden passages costumed actors take when passing unnoticed from their dressing rooms to various attractions in a theme park. If, for instance, in a hypothetical "Optics World Theme Park," the superhero Lois Laser and her sidekick Harry Hologram need to dress up in their flashy suits, emerge from the Entangled Photon café for a themed lunch with children, and then change into new outfits for a choreographed battle in the Radiation Zone arena, the park might require them to walk from one event to another via a system of enclosed staff tunnels, stairways, and corridors connecting those places. Otherwise, a wee one might cry when he sees his favorite superheroes—wearing their normal street clothes and looking completely different—sauntering into a building to get changed.

Just as in Optics World you might check a schedule to calculate your odds of observing Lois and Harry at one of their events, but you probably couldn't obtain a printout of their behind-the-scenes activities and movements (unless you happened to be a law enforcement official with a court-approved search warrant), in the particle world you can tabulate the chances of a particle's having a given position, momentum, or another observable at given moments, but you can't precisely track their entire physical profiles. The latter information is privileged to denizens of Hilbert space—in other words, no one, unless an imagined "Vicky the Vector" or "Mindy the Mindful Matrix" somehow acquire artificial awareness—not human observers.

Heisenberg himself was not troubled by the question of what physically happened during a quantum jump. Any possible in-between steps made no difference to him. What mattered was the before and after, as seen in the data. As Cassidy characterized Heisenberg's "pragmatic approach":

"We don't need to know what is happening during the jump in order to succeed in obtaining results from the theory . . . If we can't measure it, it's not part of our theories. So anything we say about it is pure speculation, or even metaphysics."[17]

Compared to our mundane realm of experiences, Hilbert space is a curious kingdom indeed. Unlike conventional space, which has three dimensions, and space-time, which possesses four, it contains an unlimited number. Indeed, many quantum systems are modeled using an infinite-dimensional Hilbert space.

How can that be? If the number of dimensions is infinite, why do we witness so few in our daily lives? What would forbid us from outdoing the movie character Buckaroo Banzai (from the film *Buckaroo Banzai Across the 8th Dimension*) and exploring an infinite-dimensional realm?

The confusion arises in the various uses of the word "dimension." Mathematicians are licensed to construct whatever kinds of abstract spaces they want, even if they aren't physically accessible. In that sense, a dimension might represent simply a possible configuration, such as whether a traffic light is flashing green, amber, or red. A "Hilbert space" depicting those possibilities might have three axes—green, amber, and red—each marked zero when that respective color is not flashing and one when it is flashing. Thus, a flashing green might be represented as "(1,0,0)"—the "1" signifying that the green light is on, and the zeroes noting that the amber and red lights are off. While such a space would have only three dimensions, it is easy to imagine situations in which there are ten possible colors (or other options) leading to a ten-dimensional space.

Now imagine a Hilbert space depicting all the possible locations of an electron along an infinite straight line. Assume a box completely confines the electron, so it cannot actually explore the whole line. If each location along the line was assigned a dimension, the Hilbert space would need to have an infinite number of dimensions.

In Hilbert space, the array of vectors representing the scope of position would span those dimensions, like a fork bent so that each prong is pointed a different way. Applying the position operator to the possible electron states (using the "sandwich" method of wedging the operator between a column and a row vector) would assign a probability for the electron's position being each of those states. The probability of the electron being outside of the box (presuming the box comprises a barrier of unlimited strength) would be zero. When integrated over the full range, these positions weighed by probabilities would produce the expectation value for position. That value might, for instance, indicate that the electron's likeliest position in physical space is at the center of the box.

Although it is eminently hard to picture a Hilbert space of multiple dimensions, let alone an infinite number, the concept does offer important conceptual advantages. For one thing, transformations between possible states might be represented as rotations (in the abstract space, not real space) from one direction to another. We might best imagine that using a flat, two-dimensional graph. The *x* axis of the graph might represent one pure state—also known as an "eigenstate"—and the *y* axis another. Rotating a state vector 90 degrees, from the *x* axis to the *y* axis, like turning a dial one-quarter of the way around, would represent transforming the vector from one pure state to another. Rotating the state vector 45 degrees, on the other hand, would place it along the *x-y* diagonal. In that case, it would no longer represent a pure state, but rather an equal mixture of two pure states. In the case of electron position, for instance, it would signify that particle being equally likely to be in two different places. Hence, that angle of rotation, known as the "phase angle," offers the perfect way to transform between quantum possibilities by simply twisting dials in Hilbert space.

Another advantage of Hilbert space is that being unlimited, it can always accommodate more dimensions. If one revises a problem by making a physical system more complicated, the Hilbert space would simply need to grow. An infinite arena can always make more space for added dynamics.

In a 1924 lecture, Hilbert cleverly described how a hotel with an infinite number of rooms, each occupied by a guest, could still make room for more. Imagine that rooms 1, 2, 3, and so forth, are each booked for a night. Unexpectedly, a limousine pulls up and a rock star emerges. She walks up to the front desk and asks the night manager if the hotel could please accommodate her. He nods, picks up a microphone, and announces on an intercom to the guests that they each need to move to the room next door. The person in room 1 should move to room 2, the person in room 2 should move to 3, and so forth. That frees up room 1 for the rock star to stay.

She thanks the night manager, but then informs him that a group of roadies that help out with the equipment are arriving soon on a tour bus and need lodging too. He asks her how many. Sheepishly, she admits that there are an infinite number. Guitar strings break, she explains, so it is best to bring unlimited backup instruments, each carried by a roadie. No problem, the manager says. He gets back on the intercom and revises his instructions to the guests. Instead of moving to the room next door, they should move to the room with double the room number. The person in room 1 should move to room 2, room 2 to room 4, room 3 to room 6, etcetera. In that fashion, all of the odd-numbered rooms would free up for the rock star and her infinite group of roadies to stay. Similarly, a Hilbert space can always make more room—even for infinitely more ways something might move. Such is the crazy logic of infinity!

The advent of quantum mechanics was jolting for those traditionalists who used physics to divide the world into two parts: things that, at least in principle, might objectively be measured, on the one hand, and intangible phenomena, on the other. The latter category included things such as consciousness, the sense of free will (even if it turned out to be illusory), ethics, aesthetics, and other abstractions that seemed hard to quantify but were universally accepted to be real, along with all manner of purported supernatural and spiritual entities, from divine beings to ghosts, that attracted some scientifically minded individuals, but certainly not all. Certainly, thanks to movements such as psychic determinism, it had become fashionable in the late nineteenth and early twentieth centuries for some thinkers to argue that eventually *everything* will find objective, mechanistic explanation.

However, the indirect process in quantum mechanics of ascertaining knowledge in a piecemeal fashion threw a wrench into that machinery. Instead of objective physical measures—the situation called "realism"—quantities are defined only through the process of measurement. Theoretically, that means applying the appropriate operator to the space of state vectors. In practice, it guarantees that nobody would be able to plot precisely the positions of, velocities of, and forces experienced by all the particles in the universe. Hence nobody would be able to map completely its web of causal connections.

Heisenberg's uncertainty principle restricts which physical parameters might be known precisely at the same time, and which must remain hazy whenever another quantity is measured. For instance, given that position and momentum cannot both be determined exactly at once, an electron of known speed has a blurred location. Similarly, because energy and time cannot both be pinned down simultaneously, and energy is related to relativistic mass, the mass of a very short-lived particle might be determined only as a range of values.

The roots of the uncertainty principle lie in the noncommutative nature of certain paired quantum operators. Because order matters in applying the position and momentum operators, each disrupts the other's ability to draw out information. For example, when the position operator is applied to a particle's quantum state, it singles out the eigenstates of position, which are generally very different from the eigenstates of momentum. Conversely, applying the momentum operator singles out the eigenstates of momentum. The clash between the two operators' actions leads to a kind of tug-of-war where the precision of one draws away from the precision of the other, often leading to inexactitude for both.

It is ironic that while relativity allowed for a strict accounting of the causal connections between events in nature, quantum mechanics subverted that process by introducing a necessary blurriness at the tiniest scales. If time is fuzzy, one cannot truly say what happens between one moment and another. A quantum jump might have a clear beginning and ending, but a hazy in-between.

Because of the uncertainty principle, knowledge of the state of the universe must necessarily have many gaps. Laplacean determinism, in which the entire future of everything in creation might be predicted indefinitely, would not just be impractical, it would be inconceivable. The cosmos is simply not a friendly place for know-it-alls; rather, like a cryptic James Joyce novel, it invites partial understanding.

Matter Waves

About half a year after Heisenberg advanced his idea of using matrices to model the probabilities of various types of discrete quantum jumps, the Austrian physicist Erwin Schrödinger independently developed an alternative called "wave mechanics" that was meant to offer a more tangible explanation of the workings of the atom. Drawing from an idea proposed by the French physicist Louis de Broglie, Schrödinger postulated that electrons are "matter waves" that obey a dynamical equation of motion—similar in spirit to Maxwell's description of light as electromagnetic waves. These quantum wave solutions, conceived as distributions of the mass and charge of an electron throughout a region of space, would become known as "wave functions." Wave functions evolve in a predictable, deterministic way according to Schrödinger's equation.

Soon thereafter, however, Born demonstrated that Schrödinger's wave functions well matched the quantum states described by Heisenberg if one equates them with probability waves, rather than matter waves. Instead of delineating an actual spread of matter, Born posited that they mapped out the chances that an electron would be located in a particular region of space. Following the notion that probability is proportional to amplitude squared, one must square the value of the wave function to get the probability distribution.

To find the exact location or speed of the electron, one must take a measurement of either one or the other. Similar to matrix mechanics, each type of measurement would be associated with a unique operator—which in this case would be a combination of mathematical operators that act on the wave function. Depending on that choice, the electron's wave function would instantly "collapse" into one of a particular spectrum of solutions of Schrödinger's equation connected with either position or momentum values, respectively—but never both at the same time.

Wave function collapse is akin to a bakery that prepares fresh loaves of bread from a homemade dough. Throughout the baking process, the ingredients and methods are so standard, no choices are made. Therefore, assuming no mistakes, it is perfectly deterministic. After the bread is ready, customers are given the choice of short slices or long slices. A slicing machine can be set up in one of two perpendicular ways: across the thin side of the bread or along its length. If the customer chooses "short," the slicer divides the entire loaf widthwise and randomly ejects one of the short slices. Similarly, it can pare out long slices, but never both short and long at the same time.

By analogy, the "slicer" of quantum measurement carves out a spectrum of Schrödinger's equation solutions, called eigenfunctions, based on the type of measurement being performed (position, velocity, and so forth). These are the equivalent to the eigenstates in the matrix mechanics approach. Each eigenfunction is associated with a certain eigenvalue, or measurement result. As soon as an experimenter takes a certain measurement, the wave function collapses randomly into a particular eigenfunction connected with a specific outcome. For a position measurement, for instance, it would collapse into position eigenfunction, associated with a certain position value.

Schrödinger was troubled by Born's reinterpretation, due to its jumpiness and abstractness. In some ways, though he had mystical tendencies and was fond of Eastern philosophy, he believed that Nature should be continuous and eminently visualizable, not fragmented and obscure. It should resemble a seamless garment, not tattered threads. According to Heisenberg, he reportedly once complained to Bohr, "If you have to have these damn *quantum jumps* then I wish I'd never started working on atomic theory."[18]

Nevertheless, as codified by other esteemed physicists such as Paul Dirac and John von Neumann, Born's fusion of the visions of Schrödinger and Heisenberg proved highly successful. Dubbed the "Copenhagen interpretation," for the city where Bohr's famous Institute for Theoretical Physics, the mecca of the field in the 1920s, was located, it became the standard view.

From that point forward quantum mechanics became a hybrid theory: a chimera with a steady, deterministic core governed by Schrödinger's wave equation and an ever-changing tangle of possible faces randomly emerging in response to what type of measurement is performed. No experimenter could possibly see all the faces at once. Whenever one emerged, such as a position value, others would hide, such as a velocity value. Quantum physics thereby embodied a fundamental incompleteness.

Because Bohr was the head of the Institute, it is commonplace to presume that his personal views matched the canon of the Copenhagen interpretation. However, as the Notre Dame philosopher of science Don Howard and others have pointed out, Bohr's views on the dichotomy between the quantum and classical world were quite distinct. Bohr's style was publically to try to pave over differences, in a kind of "big tent" approach, while privately trying to persuade others of his positions during quieter times such as long walks.

Bohr assigned classical physics two different roles, which in some ways were hard to reconcile. Firstly, in accordance with what he called the "correspondence principle," classical mechanics must match quantum mechanics for energies on the human scale—known as the "high energy limit" or "classical limit." At mundane levels, its jumpiness and uncertainty must essentially disappear. If quantum mechanics, for example, when revved up to the scale of a billiards table, predicted wrongly for the motion of colliding pool balls, then it would certainly need to be corrected. Fortunately, quantum mechanics does seamlessly match Newtonian physics (or relativistic physics for very high energies) at the classical limit. All quantum physicists would agree with Bohr, that such matching is a success.

Another role Bohr assigned to classical physics, however, would not find universal endorsement. He cordoned off classical physics as the tangible realm of experimenters, and quantum physics as a kind of "black box" hidden mechanism for churning out answers to their specific questions, depending on how such researchers construct experiments to test them.

Bohr's dividing line between the classical and the quantum connected with his advocacy of complementarity: the notion that subatomic entities, such as electrons and photons, act like either waves or particles, depending on the experimenters' intention as expressed in his or her choice of apparatus. It rendered quantum systems as somewhat like modern smartphones where, although we're eager to see the results of an application (such as a search engine or mapping program), we don't usually know or care about the underlying computations that produced them.

To return to our Optics World theme park analogy, imagine a pavilion where children get to meet their favorite characters. After making their choices, actors behind the scenes dress up in costumes and emerge from a door. To save money, the park has recruited the same two actors—one female, one male, to portray dozens of characters. From the children's perspective, they care only about the narrative of the theme park and its denizens, such as Lois Laser, Harry Hologram, and their evil twins Martha Maser and Barry Bremsstrahlung. For those who buy into that universe of characters, what goes on behind that door is simply irrelevant and immaterial. Similarly, in quantum physics, researchers focus on the results of measurements, rather than on vainly attempting to reveal the entirety of the veiled inner workings of quantum processes.

Now imagine a second pavilion where children had the opportunity to wield light sabers in the manner of their heroes. Once handed a flashlight-like device, they are given a choice: to produce a pointed beam on the opposite wall—something like the impact of a point particle, along the lines of Newton's corpuscles—or to create a striped pattern of bright and dark fringes, displaying an interference pattern due to light's wave-like nature. The device works by either sending the light through a single hole, or dividing it into two slits. The latter resembles the classic double-slit apparatus pioneered by the English physicist Thomas Young in the early nineteenth century. If the children are focused simply on pretending to be superheroes, they might not care about the mechanisms creating the particle-like or wave-like displays. Therefore, once again, they'd focus on how their choices affected outcomes in their imaginative play, rather than the "black boxes" that link the two. Similarly, we in the classical world might marvel at the results of quantum mechanisms, but must accept that a necessary, perhaps unfathomable, duality drives its inner workings.

Over the years, critics of Bohr's quantum philosophy have noted the artificial nature of his forced separation between the open, Newtonian domain of the observer and the closed, quantum enclave of the observed, especially in light of the correspondence principle that suggests their unity. In some ways (but not all), it harked back to Plato's separation between the mundane world and the realm of forms—at least in terms of the inscrutability and otherworldliness of the latter. Why should quantum systems be sphinxlike and separate except for their outcomes?

That criticism has inspired many alternatives to both Bohr's thesis as well as to the more standard Copenhagen interpretation that although it does not artificially make such a separation, it still requires an act of conscious observation to initiate a quantum measurement and a coinciding wave function collapse. Most famously, in 1957 the Princeton graduate student Hugh Everett would propose the notion of a "universal wave function" that included observers along with what is observed in a continuous system that never collapsed. He would accomplish that feat by envisioning the splitting of reality upon each quantum junction. The physicist

Bryce DeWitt popularized Everett's idea, dubbing it the "Many Worlds Interpretation." Needless to say, Bohr, by then very set in his ways, would be little interested in that renegade idea, which he essentially ignored.

Mother of Light

Despite disagreements over the philosophy of quantum mechanics, its utility is universally acknowledged. Its success in helping resolve long-standing conundrums is truly remarkable. When it yields specific answers, those are extraordinarily on the mark.

One key question that has baffled thinkers for centuries is precisely what process fuels the Sun's furnace. Before quantum mechanics and nuclear physics, no one had an adequate solution. In the nineteenth century, Lord Kelvin and the German scientist Hermann von Helmholtz each hypothesized that the Sun derived its power through shrinkage—but that would have made it far younger than the Earth!

As we now know, the Sun's inner workings depend on quantum leaps. In contrast with the steady, speed-of-light flow of sunlight through space, which takes around eight minutes, these quantum jumps may well be instantaneous or close to instantaneous. Such rapid-fire quantum processes are essential to keep the solar furnace burning. The Sun's nuclear dynamo, fueled by protons and other light nuclei, would run cold otherwise. Random, energy-releasing quantum mergers stoke its cauldron in a process called fusion—enabling it to be a stable, shining celestial body. Newtonian physics, with its iron procession of cause and effect, cannot explain why we bask in steady solar warmth. Quantum physics can.

Each photon produced in the Sun's fiery core—the central 25 percent of the star where temperatures blaze around 16 million degrees Celsius (29 million degrees Fahrenheit)—has similar origins. Defying the slim odds set out by classical (traditional, non-quantum) physics, photons emerge from change mergers of light atomic nuclei—commonly the fusion of two protons.

All life on Earth counts on the regularity of fusion in the solar core. The Sun's friendly glow makes it look easy—but it's not. Each pair of protons must conquer a steep hurdle to be able to merge.

Under ordinary circumstances, protons are antisocial particles. According to classical electromagnetic theory, like charges repel with a force called Coulomb repulsion that increases dramatically as they get closer. All things being equal, two protons, each with a positive charge, would not like to be breathing down each other's necks. Physicists represent such repulsion as a kind of energy barrier, similar to a mountain separating two isolated valleys. If two villagers on either side of a steep mountain tried to approach each other, as they got nearer they'd find it harder and harder to keep going. Eventually they'd have to turn back. Similarly, protons on a collision course would normally just scatter.

Nuclear physics tells us, however, that there are circumstances under which protons are able to pack closely together. Certain combinations of protons and neutrally charged neutrons are able to lump together into stable combinations. Deuterons are the simplest of such clusters. Two protons and two neutrons form another stable arrangement known alternatively as an alpha particle or helium nucleus. The cement that binds such stable amalgams of protons and neutrons is the strong nuclear interaction, a powerful force that acts over extremely short ranges. It is effective only on scales in the femtometers (10^{-15} meters) or less—approximately a billion times smaller than the dimensions of the tiniest bacteria.

Within the fiery solar core, the odds that two protons could overcome Coulomb repulsion and approach each other within the strong interaction range is about 1 in 10^{290} (one followed by 290 zeroes)—that is, if classical physics governed the merger. Quantum mechanics informs us, however, that protons are not just classical particles—hard spheres—but rather have a measure of fuzziness that enables them to worm their way through otherwise nearly impenetrable barriers. Each proton is associated with a probabilistic quantum wave function, which represents a smeared-out picture of its likely location. Hence two proton wave functions might overlap even when the particles themselves would be classically forbidden from approaching—offering the protons a chance of tunneling through the energy barrier and becoming close enough to meld via nuclear forces.

For any given pair of protons, it is still a slim roll of the dice that such a union happens. Causal rules can't predict it. Yet, given the immensity of the solar core, housing more than 10^{30} (one followed by 30 zeroes) pounds of incredibly dense hydrogen plasma, such quantum tunneling and fusion happens all the time. Roughly 10^{38} (one followed by 38 zeroes) protons are fused together each second in the core, producing floods of fiery photons as by-products.

When we think of tunneling through a barrier, we envision intermediate stages, like the steady advancement of construction workers and their rigs as they bore through a mountain. Yet in many quantum processes, such as electrons' jumps from one atomic energy level to another (a process that powers lasers, for example), transitions happen instantly (or at least quicker than anyone has been able to measure); there are no in-between steps. The quantum tunneling of protons in the act of fusion constitutes such a sudden transition.

In stark contrast to the sudden quantum tunneling and decay processes that give birth to the Sun's photons, their migration from its core to its surface is painstakingly slow. Once beyond the core they pass through a slightly less dense region called the radiative zone, where they ricochet between charged particles for hundreds of thousands of years. From there are two more outer regions, a thin interface called the intermediate zone and the outermost layer called the convection zone. The latter is a swirl of hot gases that act as a conveyor belt to convey heat from the more central regions to the surface. Far from the blasting nuclear furnace, the surface of the Sun is relatively "cool": only 5,600 degrees Celsius (about 10,000 degrees Fahrenheit). From there, solar photons are released into space.

Between the Sun and Earth stands 93 million miles of virtually empty space. It is especially sparse compared to the dense layers of the Sun. Yet, unlike quantum leaps, the photons don't cross that gap near instantly. Rather, due to the finiteness of the speed of light in a vacuum, it takes slightly more than eight minutes to make the crossing.

The behavior of sunlight, in its creation in the core and radiation through space, raises a fascinating question. Why does nature divide itself into quantum processes that in some cases take no discernable time, versus causal processes that have strict limitations? Part of the answer is Heisenberg's uncertainty principle, which allows for many loopholes in natural law—for example, violations of conservation of energy for very brief periods of time.

Another piece of the puzzle relates to the notion of symmetry and conservation laws. Like a zealous debt collector, reality's enforcers will travel far and wide to make sure nothing is owed in terms of conserved quantities such as charge and other particle properties. You can't cheat conservation laws by hiding. Thus, in the process of maintaining balance, certain entangled processes in quantum physics in which certain quantum numbers are shared in a multi-particle state allow for correlations over unexpectedly large distances.

6

THE POWER OF SYMMETRY

Connections Beyond Causality

What has impressed me most in the development, which in 1927 eventually led to the establishment of present wave mechanics, is the fact that real pairs of opposites, like particle versus wave, or position versus momentum, or energy versus time, exists in physics. Their contrast can only be overcome in a symmetrical way. This means that one member of the pair is never eliminated in favor of the other, but both are taken over into a new kind of physical law which expresses properly the complementary character of the contrast.

—Wolfgang Pauli, "Matter" (in *Man's Right to Knowledge*, 1954)

IMAGINE IF AN ALIEN, WHO HAS NEVER SEEN A MIRROR BEFORE, comes to Earth and first encounters one. The extraterrestrial might be astounded by the sudden presence of an identical being, or alternatively think it is a replication machine and be amazed that it works so quickly. However, there are more ways to be connected than sheer coincidence or strict mechanistic causality. Symmetry—any mathematical relationship that maps one entity onto another in a fashion that preserves particular characteristics—is another option. In principle, it might lead to instantaneous, acausal connections that could potentially connect entities that are indefinitely remote.

Of course, mirrors operate on the basis of light, and are not strictly instantaneous. However, in quantum physics there are properties that are either coordinated or anticoordinated (must be opposites). A prime example is the fourth quantum number of electrons in atoms, known as "spin," which can take on one of two possible values "up" or "down." According to Pauli's exclusion principle, no two electrons might reside in exactly the same quantum state. Therefore, a pair of electrons possessing the same energy level and orbital angular momentum quantum numbers must have opposite spin states. In other words, they are anticoordinated in spin. (Such is revealed only upon measurement; until then they are each in a superposition of the two possibilities.) Even after greatly separating the pair, if one is measured to be in the "down" state, the other must be in the "up" state, and the converse—a situation called "entanglement." That remote anticoordination of spin directions offers something like a quantum seesaw.

Binary properties, such as electron spin, show how knowledge of certain properties might be gleaned without recourse to causality. In an anticoordinated pair, revealing one member instantly tells you what the other one must be. For example, imagine if a candy factory produced jelly beans in packets of two: one red and the other blue. If you opened up a packet and the red one dropped out, you would immediately know, without even looking, that the other one must be blue. To make that determination, no information would need to be exchanged between the jelly beans. Hence, causal communication would not be involved.

Quantum physics includes many types of symmetries—some binary or another finite set of values, others with a continuous range. As we know, thanks to the brilliant research of the mathematician Amalie "Emmy" Noether, each symmetry is associated with a conserved quantity. In turn, these might offer long-range correlations—that is, remote connections without causality.

Reflections on Reflection

Mirror image symmetry, known in physics as "parity conservation symmetry," is one of the most familiar symmetries on Earth. Physically, it stems from the law of reflection—the angle of incidence of a light ray is precisely equal to its angle of reflection. For a plane mirror, that leads to an image of the same size, same vertical orientation, but opposite chirality along the horizontal direction. Chirality, or handedness, represents a division in nature between "left-handed" and "right-handed" items that are otherwise identical—such as paired gloves, paired shoes, the two sides of an arch, and so forth. Another form of chirality is clock direction—clockwise or counterclockwise. That's why screws and sink handles (to adjust the water flow), for example, come in two different versions that twist in opposite ways.

For planar objects, "living" in two dimensions, only a flip within the third dimension (at right angles to the plane) might transform a left-handed handprint into a right-handed handprint, and a clockwise spiral into a counterclockwise spiral. Three-dimensional bodies don't have the same opportunity, unless an accessible, spatial fourth dimension actually exists. Barring such access, a glove-manufacturing company couldn't save time by making only the left-handed variety and flipping half of them in the end to right-handed.

A strict form of reflection symmetry involves objects that when reflected, look precisely identical to the original versions. The letters "X" and "O" obey such a symmetry, as they look the same in the mirror. Such objects have a vertical axis of symmetry that runs through their centers, making the right side a seeming reflection of the left. Hence, a mirror would simply map left onto right and right onto left in such a way that the image preserves the original figure.

Another common symmetry is translational symmetry, which involves identical copies along a straight line. Imagine a tiled bathroom, with rows of identical squares extending everywhere along the horizontal and vertical directions. Suppose one of the tiles cracks and needs a replacement. In ordering a new one from a tiler, only the size and shape must be specified, not the location. That's because, thanks to horizontal and vertical translational symmetries, all places along the walls of that bathroom are equivalent. The need for a corner or edge tile that might not be perfectly square would "break" that symmetry (render it no longer perfect).

Rotational symmetry represents yet another familiar possibility. Imagine taking a clockface without the numerals but only twelve identical markings indicating the hours and rotating it by a quarter turn, a third turn, a half turn, or a full turn (that is, angles corresponding to an exact number of hours). Clearly, it would look exactly the same. Spin it once more in a similar fashion, and it would appear the same again. Its visage is thereby rotationally symmetric.

Many phenomena in nature emanate from a central hub, or, conversely, are attracted to a central hub. In that case, concentric circles focused on that hub often exhibit a form of rotational symmetry. Ferris wheels and merry-go-rounds offer tangible examples of that form of symmetry. Each of those carnival rides looks essentially the same as it spins.

Less obvious, though exceedingly common, examples of rotational symmetry are cases of central forces, such as the mutual gravitational pull between the Earth and the Moon. Because the Earth's attraction emanates in the inverse radial direction, the angular position of orbiting objects essentially doesn't matter for factors such as overall energy. While the Moon's orbit is not exactly a perfect circle, it is fairly close. Therefore, because the sky angle doesn't make much of a difference, the Moon's orbit approximates rotational symmetry. (The Moon's phases are patterns of light and shadow due to its position relative to the Sun's direct rays, which is a different story.)

Other common symmetries in physics include charge conjugation, which involves inverting the electric charges of two particles from positive to negative, or the converse, and noting that their attractive or repulsive behavior remains the same nonetheless. Pairs of two positive or two negative charges, assuming magnitudes and their mutual distances are the same, repel each other in exactly the same way. Time-reversal symmetry, applied to a wide variety of particle interactions, shows that forward- and backward-in-time versions appear identical—much like running a video backward of a certain process and not even noticing. And that list is far from complete. Natural symmetries truly are everywhere.

Conservation Proclamations

Nothing makes physics calculations easier than establishing a conservation principle. For example, conservation of mechanical energy guides how roller coasters would work if they were on frictionless tracks—an ideal situation, to be sure. It mandates that the cars' total mechanical energy must be recycled, not wasted. Descending a hill, each car would transform potential energy (energy of position) into kinetic energy (energy of motion). Then, climbing up the next hill the kinetic energy would transform back into potential, potentially enabling the car to ascend at least as high as the first hill. Consequently, the car's behavior would be completely predictable.

Under more realistic circumstances, the car would experience friction as it rolled along the tracks. Consequently, some of the mechanical energy would be lost, and the car couldn't climb as high on the second hill. However, if one felt the tracks afterward, they might be warmer, indicating a loss of heat—another form of energy. A careful accounting of all the waste energy expelled as heat, along with the mechanical energy throughout the ride, would show that the overall energy—mechanical plus waste—would still be conserved.

Finally, replace the roller coaster with the circular track of the Large Hadron Collider's main ring, and the cars with subatomic particles. Because those particles would typically be accelerated close to the speed of light, they would accrue relativistic mass. Following Einstein's famous energy-mass equivalence, that added mass would correspond to another form of energy. Hence the total relativistic energy (including kinetic as well as relativistic mass) would still be conserved.

An exception to the conservation-of-energy rule occurs on the tiniest level in the case of so-called quantum fluctuations of the vacuum. In accordance with Heisenberg's uncertainty principle, if a particle exists for only a fleeting life span—allowing for precise knowledge of its time interval, its energy content becomes hazier, and it might briefly break the conservation of energy law. Thus, particles might spontaneously arise from the void, live for an extremely short period, and merge back into emptiness, like a fish jumping briefly out of raging water before returning to the stream.

Normally, when charged particles emerge in vacuum fluctuations, they appear in pairs—one positive, the other negative. For example, an electron might rise up with a positron companion. One of the reasons that happens is due to another type of preservation principle: conservation of (electric) charge. A positive charge simply can't arise on its own without engendering a negative charge.

Additional conservation laws commonly used in physics calculations include conservation of linear momentum (mass times velocity) and conservation of angular momentum (essentially, mass times velocity times the radial distance to the axis of rotation). Conservation of linear momentum shows that a rocket expelling fuel in one direction (say downward) would generally move in the opposite direction (say upward) unless acted on by a net external force (a collision with space debris, for example). The downward momentum of the fuel would need to be balanced by an upward momentum of the rocket to maintain its net total. Angular momentum conservation applies more directly to circular, rather than linear, motion. As mentioned previously, it allows for stable orbits.

Remarkably, conservation laws can lead to long-distance correlations without direct causation. For example, imagine a spaceship one hundred miles long, traveling through deep space. Suppose it is a region that is so empty, any distortions due to the gravitational influence of other bodies are undetectable. Lacking external forces, the spaceship's linear momentum would be conserved. That implies that its speed would remain constant. Consequently, at least in principle, if one measured the velocity of one end of the ship, the other would instantly be known. It would represent an acausal way of immediately gleaning knowledge about a remote location. Note that in practice, however, the ship would be made of myriad atoms, each vibrating in random ways due to thermal processes. Such a system would not maintain constant speed throughout. Only if it were cooled to absolute zero—an impossibility—would it offer perfect correlations from one end to the other. Or, alternatively, if it were made of a material that exhibited quantum coherence—a lock-step inclusion of all components into a single, invariant quantum state—would it be absolutely rigid. Such materials that display quantum coherence and flow in unison include superconductors and superfluids.

As it turns out, symmetry and conservation laws always go hand in hand. Each symmetry leads to a type of conservation. In a theorem of

Emmy Noether (1882–1935), photographed near Bryn Mawr College where she was a mathematics professor; Credit: Courtesy of the Bryn Mawr College Library.

monumental importance to modern physics, the brilliant mathematician Amalie “Emmy” Noether cemented that deep connection.

Noether was born in Erlangen, Germany, on March 23, 1882, into a distinguished German-Jewish family. Her father, Max Noether, was a respected mathematics professor at the University of Erlangen. Her mother, Ida Kaufmann Noether, was an avid pianist and member of a prominent Cologne mercantile family. One of Emmy’s younger brothers, Fritz, would become a noted mathematician himself.

Noether grew up in the waning years of an era when women were excluded from academic positions in Germany and much of the world. To become a university professor—or even *Privatdozent*, the equivalent of a lecturer—was a multistep process that included achieving a doctorate and then obtaining a *Habilitation* (a higher-level degree required for university teaching). One of the first women in Germany to obtain a PhD in any subject, which she received in mathematics at the University of Göttingen, she was barred from the *Habilitation* nevertheless. Consequently, for much of her early years she relied on male professors at Göttingen, such as Hilbert, to be the official instructors of courses she taught. Appallingly, she'd be listed simply as the "assistant."

Hilbert was irate that such an impressive mind as Emmy Noether's was discriminated against because of her gender. At one point, during a faculty meeting, he blurted out, "I do not see that the sex of the candidate is an argument against her admission as a Privatdozent. After all, we are a university and not a bathing establishment."[1]

Fortunately, in 1919, within a few years after Hilbert made that remark, German regulations changed and Noether was able to complete her Habilitation. The Göttingen mathematics department approved her becoming a *Privatdozent*. Finally, she was able to teach under her own name.

It was high time for Noether to have an official title, given her rising reputation among mathematicians. In the year before achieving her Habilitation, Noether had completed a groundbreaking paper on symmetry and conservation laws. It was presented on July 16, 1918, to Göttingen's Royal Society of Science. Because Noether, as a woman, was forbidden from membership, her colleague, the esteemed mathematician Felix Klein, had to submit the paper officially. The paper would have a profound influence on the future of modern physics.

One of the impetuses Noether had for proving theorems connecting symmetry with invariant quantities was helping Hilbert complete his mathematical analysis of the general theory of relativity. Though far less famous in his contributions than Einstein, Hilbert was a major player in unraveling the overall properties of general relativity, such as how it deals with gravitational energy. A key question he wished to address was whether or not such energy, which is hard to define locally, is strictly conserved.

Noether's (main) theorem showed that symmetries imply conservation laws. It helped Hilbert establish that general relativity, which included various symmetries, implied that certain invariant quantities exist, including those related to energy. Consequently, it must conserve gravitational energy overall—even if such a quantity might not be defined from point to point.

Given the dazzling array of symmetries in nature, Noether's theorem has manifold applications. Take, for example, a spinning bicycle wheel with rotational symmetry. Such symmetry implies that angular momentum is conserved. Because of conservation of angular momentum, the motion of a bicycle wheel is more stable, as it resists the tilting of its axis without compensating by turning the wheel. Conversely, if one is riding a bicycle and does want to turn the wheel to round a corner, one might balance that turning by leaning slightly and tilting the axis of rotation. In either case, the circular symmetry leads to an invariant—an unchangeable quantity—namely, overall angular momentum.

Linear momentum, in contrast, is conserved when a different kind of symmetry is present: translational symmetry, or sameness from point to point. For example, if one plays a game of billiards on a flat, frictionless surface, there is nothing to distinguish one part of the table from another. Therefore, each ball must roll in a straight line at a constant speed, until colliding, perhaps, with another ball, or with one of the walls. In the case of such collisions, conservation of linear momentum governs how the final speeds and angles depend on the original speeds and angles.

Despite the colossal importance of Noether's theorem, she would continue to encounter major hurdles in her life. When the Nazis came into power in 1933, she would be dismissed from her position at Göttingen due to her Jewish background and socialist leanings. She managed to obtain a position at Bryn Mawr College in the United States, where she was beloved as a teacher. Sadly, though, in 1935 she'd need a medical procedure for tumors. Unexpectedly, she would die a few days after the procedure. Einstein honored her with a moving eulogy in the *New York Times*:

> In the judgement of the most competent living mathematicians, Fräulein Noether was the most significant creative mathematical genius thus far produced since the higher education of women began. In the realm of algebra, in which the most gifted mathematicians have been busy for centuries, she discovered methods which have proved of enormous importance in the development of the present-day generation of younger mathematicians. Pure mathematics, is in its way, the poetry of logical ideas.[2]

Noether's theorem offers the prospect of powerful long-range connections without causal mechanisms. Any symmetry in nature leads to an invariance: a physical quantity that remains constant until that symmetry is broken. Once a substance, or energy field, possesses such an invariance, the relevant aspect becomes as rigid as a bar of steel. It takes no causal mechanism to keep it that way. Rather, something must happen for it to break apart. In a sense, such invariances generalize Newton's concept of inertia to allow for effects without a need for forces.

For life on Earth to flourish, atoms must have a degree of stability. If atomic levels were all fragile, and electrons simply dropped on their own down to the nuclei, everything would crumble. The bulk of our bodies, for example, are stable empty spaces, supported by the solid framework of atomic and molecular structure. If those constituents collapsed, we'd no longer be here.

Luckily, the atom is chock-full of symmetries—which produce invariances leading to stability. In solving the Schrödinger equation to find the billowing probabilistic clouds of wave-function solutions, one notices many regularities. All stationary states (stable solutions that don't vary from moment to moment) have time translation symmetry, meaning that they don't change with time. The ground state (lowest energy level), resembling a perfectly round bubble, also has spherical symmetry. Other wave functions have axial symmetries, meaning that they look the same when rotated about an axis.

Following Noether's theorem, the regularities ensure a set of invariant properties, reflected in the quantum numbers. Time translation symmetry ensures that the principal quantum number, connected with overall energy, remains invariant. Therefore, for stationary states, total energy is conserved. Until that symmetry is broken, for example via interaction with an outside agent, the energy level of an atomic state remains stable. Similarly, the spherical and axial symmetries pertain to the conservation of angular momentum, represented by two additional quantum numbers. These also produce stability, until, once again, those symmetries are broken.

The stability of atoms, due to conserved quantities expressed as quantum numbers, exhibits itself in the "seating arrangements" for their electrons. Electrons group into ensembles, known as "shells," that when complete lead to greater stability. These shell patterns are akin to something like stadium seating, with fewer spaces in the inner rows and more in the outer. Electrons fill up the inner positions first, until the first shell is complete, then move on to the next shell, and so forth. Notably, the atoms in which every shell is complete are the most chemically stable. These are the noble gases, also known as the inert gases, such as helium (with 2 electrons), neon (with 10), argon (with 18), krypton (with 36), and so forth. Such stable atoms confidently stand on their own—reluctant to bond with others into molecules. Other, more active, elements, such as metals, similarly group together in the periodic table. What they have in common are incomplete outermost shells—empty seats in the "nosebleed sections" of their atomic stadiums. Hence the periodic table shows how regular patterns in the ticker-tape succession of quantum numbers, derived from the symmetry properties of atomic states, lead to the rich array of chemical element features.

Notably, though, atomic physicists noticed a glaring omission in how the quantum numbers explain the shell model. There seemed to be a missing factor of two. Start with the "magic numbers" of the noble gases, corresponding to filled outermost electron shells: 2, 10, 18, 36, and so forth. Subtracting them in succession (including subtracting 0 from the first value) results in the sequence 2, 8, 18. Taking the difference between those values (once again subtracting 0 from the first number) yields 2, 6, 10. That final sequence is precisely the first 3 odd numbers—1, 3, 5—doubled. Why one needs to double them ultimately would lead to the discovery of a fourth quantum number: spin.

A Most Exclusive Residence

In unraveling numerical mysteries, Wolfgang Pauli turned out to be the perfect man for the job. Despite being hardheaded and demanding about other researchers' theories, he had a mystical side. One of the ways it found expression was a deep interest in numerology and symmetry. He often looked for hidden patterns that might reveal deep truths. In essence, he was a kind of Neoplatonist.

One of Pauli's role models (particularly in his later years, when he'd turn increasingly to philosophy) was Johannes Kepler. Pauli was intrigued by how Kepler applied notions of mathematical beauty (such as the Platonic solids) in his quest to unravel the laws of the universe. Like Kepler, Pauli had a penchant for identifying numerical patterns that would reveal new natural principles.

Exploring the mathematical structure of the atom roused Pauli's problem-solving skills. Inspired by the discussions of others, including his mentor Sommerfeld, he attacked the conundrum of the missing factor of two with great insight and alacrity. As he later recalled:

> The series of whole numbers 2, 8, 18, 32 . . . giving the lengths of the periods in the natural system of chemical elements, was zealously discussed in Munich, including the remark of the Swedish physicist, Rydberg, that these numbers are of the simple form $2\,n^2$, if n takes on all integer values. Sommerfeld tried especially to connect the number 8 and the number of corners of a cube.[3]

Such a pattern—with possible links to three-dimensional geometry—seemed positively Pythagorean. Did it have deep significance, or was it a numerological mirage, like Kepler's Platonic solid hypothesis? An important clue was that the three then-known quantum numbers could explain everything except the doubling. That suggested—in Pauli's mind—a fourth, two-valued quantum number.

Along with the pattern puzzle, another atomic conundrum pertained to their stability. Why didn't the electrons in the higher elements all crowd into the lowest positions? What stopped them from falling all the way down by emitting photons of the right frequencies? If minimizing energy were the only component of their "decision-making," they'd all flock to the ground state. But in that case, most atoms would be chemically unstable, including the oxygen we breathe and the iron in our blood, and we wouldn't exist. The fact of our bodies being solid structures meant that something was forcing the electrons into higher levels, like sports fans moving to the outer rows of an arena once the inner seats are all taken. An unknown agent seemed to be acting like a bouncer and guaranteeing that the outermost electrons don't crowd the inner shells of atoms.

A simple, but deceptively subtle, type of symmetry, as applied to the particle world, is exchange symmetry. Subatomic particles are generally indistinguishable (except for their quantum numbers, which pertain to their conditions, not their identities). That means that a statistical description of a heap of such constituents would not take into account their order. Switch electron A from state 1 to state 2 and electron B from state 2 to state 1 and you couldn't tell the difference.

Exchange symmetry applies, however, to the wave function of a quantum system that includes two identical particles, such as two photons. As Pauli demonstrated, such combinations have distinct properties depending on the particle type. One category, called "bosons," named for "Bose-Einstein statistics" and encompassing particles such as photons, is characterized by combinations that are symmetric under exchanging the quantum numbers associated with the particles. Therefore, identical bosons can effectively share the same quantum numbers. The other category, called "fermions," named for "Fermi-Dirac statistics" and including particles such as electrons, is characterized by combinations that are antisymmetric under such exchanges. Essentially, that means if you exchange the quantum numbers of paired fermions once, you don't get the same thing, but rather you get something like an inverse wave function. You would need to exchange quantum numbers twice to get back to the original. The difference is akin to exchanging the right and left sides of the letter "X" (representing paired bosons) or the letter "N" (representing paired fermions)—the first is symmetric and the second is antisymmetric.

Pauli proposed, in rudimentary form, what he called the "exclusion principle" even before Heisenberg and Schrödinger completed their versions of quantum mechanics and the concept of wave functions was proposed. The first known expression of it, in December 1924, sounds like an edict:

"It shall be forbidden for more than one electron [in the same atom] . . . to have the same values of [all applicable] quantum numbers."[4]

In other words, no two electrons—or fermions in general—are permitted to share exactly the same set of quantum numbers. At least one

Austrian physicist Wolfgang Pauli (1900–1958) in his youth, presenting a lecture on quantum mechanics; Credit: AIP Emilio Segrè Visual Archives.

thing must be different. Later Pauli justified applying the exclusion principle to fermions, but not to bosons, because of paired fermions' antisymmetry under the exchange operator.

Pauli further speculated that the difference between electrons that have shared the first three quantum numbers, such as the two occupants of hydrogen's ground state, must lie in the values of a fourth quantum number, which might take on one of two possible values. Believing that it was indescribable using classical physics, he refused to say what the fourth quantum number physically represented.

Pauli's exclusion principle, and notion of a new two-valued quantum number, found ready acceptance, as it offered a natural explanation for why atomic shells have the particular sets of values for the numbers of electrons that fill them, as well as a reason electrons could not crowd the lowest quantum states. The exclusion principle of one state, one electron, kept them from clustering. It created a kind of pressure for the electrons to populate the higher levels of atoms, as well as the lower. In short, it helped justify the arrangement of the periodic table that so elegantly describes the key properties of the building blocks of all things on Earth. For that significant discovery, Pauli would be honored with the 1945 Nobel Prize in Physics.

Spin: The Implausible Quantum Property

Pauli's declining to provide an explanation for his new "two-valued" quantum number proposal offered the opportunity for other eager young physicists to have a go at the question—with the aim of resolving the anomalous Zeeman effect and explaining why certain atomic numbers offer more stability. Recall that the Zeeman effect is when a magnet acts as a kind of "prism" and splits single-hued spectral lines into a rainbow of hues. The "anomalous" variation applies to atoms with an odd number of electrons, unexpectedly yielding an even number of spectral lines in the resultant rainbow.

One such innovative thinker was the German physicist Ralph Kronig. While working in the laboratory of Alfred Landé at the University of Tübingen, he met Pauli and suggested the idea that electrons rotate about their axes like spinning tops, acting like miniature electromagnets, and thereby generating an interaction with external magnetic fields. The fourth quantum number would thereby pertain to the intrinsic angular momentum associated with such rotation. Pauli immediately dismissed the notion as "unnatural," given that there was no direct evidence for such twirling. In light of Pauli's opinion, Kronig never published his idea.

As Kronig later recalled:

"I think the idea that this electron should turn about its axis was especially unsympathetic to Pauli. He didn't want to have a model for the fourth quantum number."[5]

Blossoms might sprout in furtive beauty before they meet an appreciative eye. In the case of the concept of electron "spin," as the property became named, it took a second, independent flowering of the concept before it was ripe. Luckily in the second case, Pauli never had the chance to nip it in the bud.

In spring 1925, two young Dutch physicists, Samuel Goudsmit and George Uhlenbeck, were working together at the University of Leiden on interpreting atomic spectra and trying to explain the anomalous Zeeman effect. After learning about Pauli's hypothesis of a fourth quantum number, Goudsmit explained the reasoning to Uhlenbeck. In turn, Uhlenbeck suggested the concept of a spinning electron interacting with an external magnetic field. If the electron were rotating counterclockwise, its spin axis would point up. If the external field were also pointing in that direction, the two would be aligned. If, on the other hand, the electron were rotating clockwise, its spin axis would point down, and, compared to the external field, it would be counteraligned. Whether there was an alignment or counteralignment would offer a slight difference in the electron's overall energy. Added to other factors that pertain to energy—such as orbital distance, shape, and orientation—it would produce the subtle frequency effects seen by atomic spectroscopists.

The fourth quantum number, spin, would have one of two possible values. "Spin up" would correspond to positive one-half times h-bar (Planck's constant divided by two pi), and "spin down" would match negative one-half times h-bar. They were the first non-integer quantum numbers. In line with the exclusion principle, the lowest shell of a multi-electron atom such as helium would include one electron with "spin up" and another with "spin down." Only when an external magnetic field is turned on would such states have different energy levels.

The two researchers wrote up the idea and presented it to their research supervisor, Paul Ehrenfest, with the aim of having it published. Like Pauli, Ehrenfest—who similarly hailed from Vienna—was an eccentric character, brimming with dark humor and a cynical attitude. Overtly, though, Ehrenfest, who possessed a lifelong inferiority complex (and, plagued by depression, would eventually take his own life), was much more deferential to other physicists—many of whom he invited to Leiden for seminars. Rather than dismissing others, he would bombard them with targeted questions. An emblem of how the physics community viewed the two was a parody of Faust, enacted by young physicists at Bohr's Institute in 1932, in which Pauli was depicted as Mephistopheles (the Devil) and Ehrenfest as the troubled, self-doubting titular character.

After handing Ehrenfest their work, Goudsmit and Uhlenbeck continued to investigate its implications. They shared their work with Lorentz, who had stepped down from his theoretical chair at Leiden but remained interested in contemporary physics. He pointed out a disturbing fact about their model. To match the rotational rate needed to create the requisite magnetic interaction, electrons would have to spin such that points on their surfaces whirl far, far faster than the speed of light. That meant their notion was embarrassingly unphysical.

Uhlenbeck approached Ehrenfest and told him that they had changed their minds, and that their paper was not yet ready for publication. "It is too late," Ehrenfest replied. "I have already submitted the paper. It will be published in two weeks." But all wasn't lost, he counseled them. "Both of you are young and can afford to do something stupid."[6]

Hence Goudsmit and Uhlenbeck's quantum spin proposal was published, before Pauli or anyone else could raise objections. Predictably, Pauli criticized the concept once he saw it in print. However, he'd soon change his mind and be supportive. What swayed him was a phenomenon known as "Thomas precession," discovered by the British physicist Llewellyn Thomas, that showed how the interaction between spin and special relativity could explain certain spectral properties.

While spin caught on because of its predictive success, it is not a simple concept to wrap one's mind around. Because of the speed-of-light limit, electrons can't actually rotate that fast. Rather, their intrinsic interactions with magnetic fields mimic the effects of spinning tops, without any actual twirling going on. Hence the term "spin" has stuck to describe one of the fundamental quantum properties.

Electrons aren't the only particles with spin. Quarks, and all other fermions, also have half-integer spin. In the case of two fermions sharing the same energy level (and other properties), if one is "spin up," the other must be "spin down," and the converse. Photons, and all other bosons, in contrast, have integer spin. For example, a photon has spin one (times h-bar).

The spin states of photons are usually called polarization states, either counterclockwise (spin of positive one times h-bar) or clockwise (spin of negative one times h-bar). Polarization offers a prime example of a quantum property connected with a familiar experience. Put on a pair of polarizing sunglasses, and glare is significantly reduced. That's because thin molecules in the material allow only half of the polarization modes to get through, while blocking the other half, thus reducing the brightness by 50 percent.

Total angular momentum—which includes both spin and orbital angular momentum—is a conserved quantity. Therefore, particles might change their spin states only by converting it somehow into another type of angular momentum or exchanging its quantity via another particle. Such conservation offers a further reason why if a pair of electrons stem from the same atomic level, implying that their spin state is an equal superposition of spin up and spin down, that balance must be preserved even if the electrons become widely separated. In general, conservation laws, with their ironclad system of credits and debits that spans all of nature (as far as we can tell), are able to produce nonlocal effects—like a debt collector traveling far and wide to gather payments.

The Stealth Particle

One remarkable thing about conservation laws is how they are able to help identify missing natural ingredients. Just as our ledger balances might lead us to conclude that we somehow forgot to list a source of income or type of expense, they tell us which components of an interaction might have escaped notice.

In studying radioactive nuclear decay, physicists observed situations in which atomic nuclei would emit high-energy electrons, called (for historical reasons) beta particles, in a process known as "beta decay." In the process, the nuclei would become more positively charged by the amount of a proton, moving up one notch on the periodic table and becoming an isotope of a higher element. After carefully measuring the energy and momentum of the initial and final components of the decay, several mysteries arose. Some unseen agent seemed to be carrying away a measure of the energy and momentum. Photons could not explain the situation, as they'd readily be detectable. Moreover, each time a nuclear decay produced an electron, it was generating, seemingly out of the blue, a new spin ½ particle. Surely, spin couldn't just come out of nowhere.

In 1930, Pauli—who by then was professor of physics at the ETH (Swiss Federal Institute of Technology) in Zürich—offered a brilliant explanation: a new, lightweight, neutral particle of spin ½, which he tagged the "neutron." In 1932, the Italian physicist Enrico Fermi would re-dub it the "neutrino" after James Chadwick discovered a more-massive neutral particle, also designated the neutron. Pauli announced his original idea in a particularly dramatic way to the attendees of a December 1930 research conference in Tübingen, by sending them an amusing letter.

> Dear Radioactive Ladies and Gentlemen . . .
>
> I have hit upon a desperate remedy to save the "exchange theorem" of statistics and the law of conservation of energy. Namely, the possibility that there could exist in the nuclei electrically neutral particles, that I wish to call neutrons, which have spin ½ and obey the exclusion principle and which further differ from light quanta in that they do not travel with the velocity of light. The mass of the neutrons should be of the same order of magnitude as the electron mass and in any event not larger than 0.01 proton masses. The continuous beta spectrum would then become understandable by the assumption that in beta decay a neutron is emitted in addition to the electron such that the sum of the energies of the neutron and the electron is constant . . .
>
> Your humble servant,
> W. Pauli[7]

As we've noted in our discussion of the faster-than-light neutrino hypothesis, neutrinos have been the subject of geeky humor. The tradition of poking fun at neutrinos, as it turns out, dates back to Pauli's proposal. For example, in the 1932 Copenhagen parody of *Faust*, the Gretchen character, whom Faust, portrayed as Ehrenfest, was driven by Mephistopheles, depicted as Pauli, to seduce, was identified as a neutrino. The implication was that Pauli could instill passion for a mere theoretical construct.

Another depiction of the neutrino as an attractive woman was in the poem "*La Plainte du Neutrino*" ("The Neutrino's Complaint") penned by the quantum physicist Léon Rosenfeld for the *Journal of Jocular Physics* as a parody of "*Un Secret*" by the French poet Félix Arvers. The satirical version emphasized the neutrino's enigmatic nature, like a mysterious lover. For example, Rosenfeld replaced the final line of the original poem, "'*Quelle est donc cette femme?' et ne comprendra pas*" ("'What then is this woman?' and not comprehending"), with "'*Quelle est cette énergie?' et ne comprendra pas*" ("'What is this energy?' and not comprehending"), thereby conveying how hard it would be to capture such an elusive particle.[8]

Shortly after Chadwick's discovery of neutrons, Fermi identified mechanisms by which beta decay could transpire and calculated their transition probabilities. A neutron could decay into a proton, an electron, and an antineutrino (the antimatter counterpart of neutrinos). Or conversely, a proton could transform into a neutron, a positron, and a neutrino. He worked out a formula dubbed "Fermi's Golden Rule" to calculate the odds of each transition.

Fermi trusted that Pauli's hypothesis was correct, despite the lack of experimental evidence for neutrinos. Being very lightweight (Fermi assumed massless) and electrically neutral, no one thought they'd be found very easily. They could seamlessly pass through the entire Earth as if it were tissue paper.

Though innovative, Fermi's model had several major gaps. Firstly, he did not adequately explain the conduit for beta decay. Photons mediate the electromagnetic interaction. Like charges repel and opposites attract because of photons exchanged between the two of them. Decades later the physicists Steven Weinberg, Abdus Salam, and Sheldon Glashow would extend Fermi's model into a full theory of the electroweak interaction: a combination of electromagnetism and the weak interaction. The weak interaction, in turn, generalizes the beta decay process to encompass a wide range of decay transformations involving multiple particles. It is mediated by one of three exchange bosons, the W+ and W- (involved in the two different kinds of beta decays), and the Z^0 (a neutral weak exchange).

Another absence from Fermi's model, and Pauli's hypothesis that inspired it, was the confirmation of actual neutrinos. It is one thing to propose a new natural mechanism; it is another to check that all its components match expectations. While the versatile Fermi was adept in laboratory settings, his interests soon turned to other kinds of nuclear processes. During World War II, he would pioneer the chain reaction in nuclear fission.

Pauli, on the other hand, would hardly be invited to a lab to test his own hypotheses. Experiments and he did not mix. On the contrary, in what was dubbed the "Pauli effect," he was known to disrupt equipment in whatever experimental facility he entered, or even walked near. Machinery would break down, instruments would fail, and mayhem would ensue. At least one researcher, Otto Stern, wanted to ban Pauli from passing through the doors of his institute. As the physicist George Gamow described the effect:

"The standing of a theoretical physicist is said to be measurable in terms of his ability to break delicate devices merely by touching them. By this standard, Wolfgang Pauli was a very good theoretical physicist; apparatuses would fall, break, shatter, or burn when he merely walked into a laboratory."[9]

One of the most notorious examples happened in February 1950, during one of Pauli's visits to Princeton. The powerful cyclotron (circular particle accelerator) in the basement of the Palmer Physical Laboratory caught on fire and burned for more than six hours. The entire building was stained with soot and smoke. Pauli wasn't on-site, just in the vicinity, but was blamed anyway.

Another time, Pauli happened to be passing through Göttingen. When his train stopped in the station, equipment at the university exploded for no particular reason. The effect also worked in reverse; attempts to trigger it deliberately seemed doomed to failure. At a conference in Italy that Pauli was attending, students tried to rig up a system for a chandelier to drop on him when he opened a door. The rope became stuck and nothing happened. He was amused by the counter-example.[10]

During a stay in Copenhagen during the 1950s, the physicist Stanley Deser recalled experiencing Pauli's predilection for demolition: "He nearly destroyed my then sports car, when I was delegated to take him somewhere at the Bohr Institute, the usual Pauli effect."[11]

In 1956, the neutrino would finally be discovered. The physicists Clyde Cowan and Frederick Reines detected it, helping confirm Pauli's hypothesis and aspects of Fermi's model. Along with the "electron neutrino" involved in beta decay, two other kinds of neutrinos would be found in subsequent decades: the muon neutrino and the tauon neutrino, associated with particles similar to the electron but more massive, the muon and tauon, respectively.

Electrons, muons, tauons, and the various neutrino varieties form part of a general category called "leptons": signifying particles that don't respond to the strong interaction. Conversely, protons and neutrons are "hadrons," meaning they do engage in strong interactions—which is the force that cements them into the nucleus. In the 1960s, the American physicist Murray Gell-Mann would identify the components of hadrons as subconstituents called "quarks."

An Entangled Yarn

Symmetry considerations motivated particle physics breakthroughs in many other ways. For example, the Dirac equation—developed as a variation of the Schrödinger equation for fermions that took their half-integer spins into account—offered two different solutions, seemingly of opposite energies: one positive, the other negative. Dirac interpreted the latter as positively charged "holes" in an infinite energy sea, which represented places where electrons emerged. As his construct was unwieldy, a better solution presented itself with the notion of positrons—and antimatter in general.

Antimatter turns out to be a mirror image of ordinary matter in some ways, and a carbon copy in others. The antimatter counterpart of a charged particle, such as a negatively charged electron, has the opposite charge, such as a positively charged electron. If the two encounter each other, they mutually annihilate, producing uncharged photons that carry off the energy. Physicists have defined a conserved quantity called a "lepton number" that takes on positive values for leptons, such as electrons and neutrinos, and negative values for their antimatter near-equivalents.

In electron-positron annihilation, the positive lepton number of an electron (+1) adds to the negative lepton number of a positron (–1) to yield zero, the lepton number of a photon (because it isn't a lepton).

Beta decay shows how lepton number conservation works. When a neutron decays into a proton, an electron and an antineutrino (with lepton number –1) form in tandem and balance each other out. Thus the overall lepton number remains constant.

Despite differences in lepton number and electric charge, antimatter particles bear aspects that are identical to matter—for example, their masses and the way they respond to gravity. Physicists have been able to build and isolate antimatter atoms in the lab. Antimatter is typically short-lived, however, mainly because it reacts so readily to ordinary matter.

Many symmetry ideas are best represented as rotations in an abstract Hilbert space. These are understood through the mathematics of what is called "group theory." Spin offers such an example. It might transform, through a kind of turn of the dial, from a superposition of "spin up" and "spin down" into a pure state of either "spin up" or "spin down," reflecting, perhaps, the influence of an external magnetic field.

The mechanics of spin states are best expressed as variations of vectors, called "spinors." These, in turn, might be represented as combinations of 2-by-2 numerical arrays called the Pauli spin matrices. Including the identity matrix (equivalent to the number one), there are four different types of Pauli matrices, which are closely connected with the system of four mathematical entities called "quaternions," invented by the Irish mathematician William Hamilton. They are generalizations of the complex (real and imaginary, grouped together) numbers, used in quantum mechanics and many other fields.

Another symmetry expressed through rotations in an abstract space is "isospin" (originally known as "isotopic spin"), a hypothesized near-symmetry between protons and neutrons introduced by Heisenberg in 1932. The idea is that a representation of one particle might be rotated into the other within an abstract space that included both possibilities. In what he called the "Eightfold Way," named after a set of Buddhist tenets, Gell-Mann later generalized isospin into a larger symmetry group that included more massive particles.

Principles established in nature might be altered or abolished once new evidence emerges. Just when nonlocal interactions seemed headed for the dustbin of history as a failing of Newtonian gravitation, quantum mechanics resurrected the concept—not for forces per se, but for correlated properties. Two particles might become linked remotely, such that their conditions are correlated.

Schrödinger coined the term "entanglement" to characterize such a situation. Entanglement is when a quantum state encompasses a system of two or more particles—for example, two electrons in the ground state (lowest energy level) of a helium atom—in such a way that the properties of each particle are linked to those of the others. Strangely enough, entanglement does not seem to respect physical distances. Experiment after experiment has extended the limits of how far apart the entanglement of a quantum system might reach: not just on atomic scales, but across rivers and even outer space. Far from being an abstraction, entanglement is extraordinarily useful, as it leads to organization in materials that otherwise wouldn't emerge. Examples—each operating below certain critical temperatures—include superfluids, frictionless fluids that flow absolutely smoothly, and superconductors, perfect conductors with no electrical resistance.

Resisting the Supernatural

A fundamental feature of Einsteinian physics is "local realism": a mandate that natural phenomena are guided by the objective physical conditions of their immediate surroundings. We can break that down into two concepts: locality and objectivity. Locality means no physical interaction can leap distances; rather it requires an intermediary, called a "field." As we've discussed, a field is a point-by-point map of how any force affects objects, much like a weather map might tell you the wind's speed and direction anywhere in a region. Plop a particle into a field at a given point and you know what it will do.

Contrast the locality condition with Newton's way of describing gravitation as a kind of invisible connection between massive bodies. In Newtonian physics there is nothing between the planets and the Sun that conveys the gravitational force. Rather it is simply a long-range linkage through empty space: an action at a distance. Newton was himself perplexed by that logical gap in its theory. He wanted to find an intermediary. (In fact, physicists would later rewrite Newtonian physics as a field theory—with a local gravitational field conveying the force at each point.)

Objectivity, in Einstein's view, requires that physical conditions exist prior to and independent of how they are measured. Newton's theories were clearly based on objectively determined properties, such as mass, velocity, and so forth. In that aspect, Einstein simply followed suit. He maintained that any physical theory within objectivity must be incomplete—an artifact of lack of knowledge.

A clichéd philosophical question asks what would happen if a tree fell in a forest and no one were there to hear it; would it still make a sound? Given that sound is caused by the disturbance of air molecules, Einstein would say "yes." If a woodsman discovered the fallen tree later on the forest floor, he would potentially be able to calculate how much energy was conveyed to the air molecules nearby as it fell, and estimate how much sound was made. A more detailed computer model might potentially be able to track the positions and velocities of all the displaced air molecules at any point from the instant of falling onward.

Einstein's masterful general theory of relativity certainly obeys local realism. It asserts that each event in space-time is essentially its own world, shaped by its immediate, objective conditions—namely the "shape" of space-time itself in that region. Only by linking each patch with the next in a kind of cosmic quilt does the overall pattern of the universe emerge.

For example, consider why the Earth revolves around the Sun. In contrast to the Newtonian view that gravitation links the two bodies like an invisible cord, Einstein posited that space-time in the solar system is locally warped by the Sun's massive presence. The amount of warping at each point objectively depends on the Sun's mass and energy. If space-time were flat, the Earth would move in a perfectly straight line. However, because space-time is curved, Earth's path is bent. Like a bicycle in a velodrome, the Earth is compelled to follow a closed orbit. In short, local, objective conditions steer the Earth around the Sun.

For quantum physicists, entanglement is a legitimate correlation resulting from two objects, however distant, sharing the same quantum state. However, Einstein considered the knowledge by one particle of another's condition tantamount to telepathic communication. Having grown up during a time when many scientists associated psychic phenomena with hucksters such as Slade (despite a few such as Zöllner who took his work at face value), Einstein had a marked skepticism for anything that smacked of "mind reading."

As the Caltech historian of science Diana Kormos-Buchwald, director and general editor of the Einstein Papers Project, has noted, to her knowledge none of his collected documents "support any form of psychic phenomena or mysticism."[12]

Nonetheless, during visits to Southern California in the early 1930s, Einstein faced surprising requests to endorse the purported feats of "mind readers." Because of his magnanimity, he remained gentle in his critiques of alleged psychics. Accordingly, his polite disagreement was occasionally misconstrued as friendly support.

For example, consider the time in 1930 when the prominent author Upton Sinclair (*The Jungle*), wrote the book *Mental Radio*, extolling the possibility of telepathy, and asked Einstein, a friend of his, for an endorsement. Sinclair's very unscientific account of how his wife seemed to have an uncanny knack for finding items he lost (among other feats) seemed an unlikely treatise for Einstein to recommend. Yet, apparently as a favor, he did. In his preface to *Mental Radio*, Einstein wrote:

> The results of the telepathic experiments carefully and plainly set forth in this book stand surely far beyond those which a nature investigator holds to be thinkable. On the other hand, it is out of the question in the case of so conscientious an observer and writer as Upton Sinclair that he is carrying on a conscious deception of the reading world; his good faith and dependability are not to be doubted. So if somehow the facts here set forth rest not upon telepathy, but upon some unconscious hypnotic influence from person to person, this also would be of high psychological interest.[13]

Then in March 1932, the *New Republic* featured a story about Einstein's purported endorsement of the controversial self-proclaimed clairvoyant Gene Dennis. After meeting Einstein in Palm Springs, California, likely through a shared automobile ride when he was vacationing at that resort, she alleged that he recognized her ability to foretell the future. Once again, he was probably just being polite. "Why, Dr. Einstein!" was the headline of the article about their supposed connection. Sinclair leapt to Einstein's defense, in a way that he likely didn't welcome, by falsely explaining that "Professor Einstein has long been concerned with psychic matters and has done some investigation in the field."[14]

It is conceivable that Einstein's disdain for any association with so-called psychic phenomena hardened even further his opposition to the notion of long-range, acausal connections in quantum mechanics. He found himself on a singular crusade to make all linkages clear and explicit. Ultimately, his insistence on an objective reality independent of observation would marginalize him in the physics community.

7

THE ROAD TO SYNCHRONICITY

The Jung–Pauli Dialogue

> *Professor Einstein was my guest on several occasions at dinner . . . These were very early days when Einstein was developing his first theory of relativity . . . It was above all the simplicity and directness of his genius as a thinker that impressed me mightily and exerted a lasting influence on my own intellectual work. It was Einstein who started me thinking about a possible relativity of time as well as space, and their psychic conditionality. More than thirty years later, this stimulus led to my relation with the physicist Professor W. Pauli and to my very thesis of psychic synchronicity.*
>
> —Carl Jung (letter to Carl Seelig, February 25, 1953)

EINSTEIN CONSIDERED THE ABOLITION OF NEWTONIAN "ACTION AT a distance" one of his principal accomplishments. By mandating a speed limit for communication, relativity eliminates many tricky situations, such as the prospect of a planet's instantly changing its motion if its star explodes, before any light from that explosion has a chance to reach it. Shouldn't particles, Einstein thought, behave the same way as astral bodies, with objectively defined characteristics that change only through causal actions conveyed from point to point through space?

Quantum mechanics offered correct predictions, Einstein concluded, but was incomplete without local, objective specification of its measurable properties.

A more comprehensive theory was needed, Einstein believed, that somehow allowed for all physical parameters to have values before they are measured. These values must depend tightly on other quantities in their vicinity, like the interlocking gears and chains driving a bicycle. The possibility of furtive cables of causality driving quantum entanglement from behind the scenes has been dubbed "hidden variables."

While Pauli shared many interests with Einstein, including the pursuit of unified theories of nature, he maintained a starkly different attitude toward quantum mechanics. Novel quantum phenomena, such as complementarity, the uncertainty principle, and entanglement, Pauli felt, represented the true face of reality. They offered an ideal opportunity to fathom how nature is interconnected by means of symmetries and other mathematical relationships. Furthermore, by placing the observer in a central role, quantum measurement theory hinted at the possibility of a grand vision of everything that encompassed consciousness along with physics. In Jung, who was fascinated by modern physics and appreciated it on a colloquial level, he would find the perfect intellectual partner for discussions about possible links between mind and matter. That would lead him to wonderful scientific explorations—of, for example, the role of symmetry in the universe—as well as weird excursions into the realm of so-called psychic phenomena. Pauli didn't generally advertise his interest in the latter to other physicists (except for his friend Pascual Jordan, who shared such pursuits). Knowing Einstein's stubbornness about objectivity in Nature, Pauli was even more cautious. He didn't try to press Einstein to consider the role of observers in physics. As he knew, when it came to accepting the strange implications of quantum mechanics, no one could sway Einstein.

Disentangling Entanglement

Troubled by the nonlocality of entanglement, as well as its seeming incompleteness (due to quantities that remain unknown until observation), in 1935, Einstein, along with his assistants Boris Podolsky and Nathan

Rosen, published an influential paper, now known as EPR. (Podolsky, as it turns out, was the actual writer.) While their paper considered conundrums involving the positions and momenta of entangled particles, another physicist, David Bohm, soon recast EPR more simply in terms of spin. In Bohm's version of EPR, two entangled electrons (energetically released from the ground state of a helium atom, for example) are sent off in opposite directions before either of their spins are measured. Upon measuring one of their spins, the spin of the other is instantly known. For example, if one is spin down, the other must be spin up. How could such information about one of the electrons become immediately accessed by the other?

To make matters even weirder, as shown in the famous Stern–Gerlach experiment, spin states are definitive only with respect to a particular axis—normally x, y, or z—that pertains to the direction of the external magnetic field used to take a measurement. If an electron is spin up or spin down with respect to the z axis, it has indefinite spin states with respect to the x and y axes. Thus, in Bohm's rendition of the EPR thought experiment, the second electron also happens to "know" the axis along which the first was measured.

Encouraged by Einstein, Bohm began to challenge the orthodox interpretation of quantum mechanics. Building on earlier work by De Broglie, Bohm explored the possibility of realistic particles guided from point to point via unseen "pilot waves" operating in the background. The dynamics of those waves, driven by a deterministic engine akin to the Schrödinger equation, would serve as the hidden variables underlying entanglement. Ultimately, though, while appreciating Bohm's independence of thinking, Einstein was not a fan of his methodology. While Bohm's construct was deterministic, it lacked local realism, which for Einstein in his later years had become even more essential. "[Bohm's] way seems too cheap to me," he wrote to Born.[1] Instead Einstein felt that extending general relativity into a full-blown unified field theory would reveal the unseen dynamics underlying quantum phenomena.

Pauli was much harsher, writing to Bohm, "This is nonsense . . . and not even new nonsense."[2] He was referring to De Broglie's earlier version that he had also criticized.

As much as he favored local theories, Einstein did speculate, however, about one possible source of nonlocality in general relativity. In 1936, along with Rosen, he conjectured that one part of space-time could be warped sufficiently by matter and energy so that it becomes connected with another, otherwise separate, space-time region. Such a linkage has become known as an Einstein–Rosen bridge or a wormhole—the latter suggesting a tunnel through a cosmic "apple," enabling a direct link through its pulp between two parts of its surface.

Until the end of his days, Einstein continued to argue that quantum mechanics was incomplete. It was an ongoing debate between Einstein and practically the rest of the physics world. Failing to convince his colleagues, he ardently hoped to tie up its loose ends by perfecting a unified field theory that extended general relativity to encompass all of nature. He hoped that such a final theory would somehow explain the quirks of quantum mechanics as mathematical oddities within an otherwise smooth theory—much like turbulent whirlpools sometimes arise naturally from steadily flowing streams.

The notion of unifying all natural phenomena by means of a single explanation arguably dates at least as far back as the Pythagorean belief that its fundamental components were numbers, specifically the first ten integers. In more modern times, Maxwell's unification of electricity and magnetism into electromagnetism was an enormous step. Maxwell himself recognized commonalities between electromagnetism and gravitation that might conceivably point the way toward a united theory—which he never personally attempted.

Shortly after Einstein published the general theory of relativity, he became aware of three attempted methods for extending it to encompass electromagnetism. The first, by Hermann Weyl (a friend and colleague of Noether's at Göttingen), imagined modifying the definition of the four-dimensional equivalent of length by means of including a variable factor called a "gauge." Although Weyl's unification was unsuccessful, he later successfully applied the gauge idea to quantum field theory—showing how energy fields could have arbitrary internal factors acting like spinning dials pointed randomly in any direction.

Another idea, proposed by Eddington, modified the way vectors are moved through curved space-time to include an extra factor related to electromagnetism. Finally, a third proposal, by the mathematician Theodor Kaluza, imagined adding a fifth, undetectable dimension to four-dimensional space-time, with the goal of opening up room for an expression of Maxwell's electromagnetic relationships.

Einstein was interested in those ideas, attempting, throughout his later career, variations of Eddington's and Kaluza's proposals. It was ironic that Einstein would explore the fifth dimension as a conduit for unification, given his critique of the lack of realism in quantum mechanics. He seemed to prefer hidden connections behind the scenes of ordinary space-time that could explain long-distance correlations rather than the complete abstraction of Hilbert space. The former still followed deterministic rules and maintained continuity, which for Einstein made a world of difference.

The Cynic and the Rebel

Pauli similarly maintained a critical fascination with unified field theories. He seemed to enjoy learning about them in order to debunk them. Einstein would use Pauli as a "sounding board" to sense his opinion about various schemes, which Pauli would inevitably tear apart. At one point Pauli mocked Einstein's penchant for issuing theory after theory:

"[I]n a variant of the well-known historical saying," Pauli wrote in a scholarly review, "one might exclaim at the appearance of a new attempt on this subject: 'The old theory of Einstein is dead. Long live the new theory of Einstein!'"[3]

In another example of Pauli's cynical involvement in the unification struggle, he brusquely told the Swedish physicist Oskar Klein, who had independently attempted to unify gravitation and electromagnetism within a five-dimensional framework, in a manner connected with quantum mechanics, about Kaluza's earlier work. He later helped convince Klein to give up the fifth dimension in favor of more conventional quantum ideas, such as the Dirac equation. One evening Pauli and Klein shared a bottle of wine and toasted to the death of the fifth dimension.[4] Soon thereafter, Pauli coldly told Klein, "I am not of the opinion that finding new laws of nature and indicating new directions is one of your great strengths."[5]

Pauli's mocking of other theorists and their ideas became notorious. He once offered the opinion that an idea was so bad that it was "not even wrong." He belittled a junior researcher as "so young and already so unknown." Recognizing his cruel reputation, he sometimes signed his letters either "*der fürchterliche Pauli*" (the horrible Pauli) or "*die Geissel Gottes*" (the scourge of God). As the quantum physicist Kurt Gottfried, who met Pauli at least once in the 1950s, recalled, "He was notorious for being difficult."[6]

The theorist Stanley Deser similarly noted, from a time they were both at the IAS (Institute for Advanced Studies) in Princeton, how Pauli would offer devastating remarks at seminars. The only salve was that everyone knew his personality and could brace themselves. As Deser recalled:

> Pauli frequently destroyed people when he was around. He would tear them apart so they would leave the seminar room in tears. Since it was all expected from him, no one took it badly—at least not permanently.
>
> [He] was a great man, but devoid of the usual polite instincts. I recently discovered an old letter from him. When I was going to Copenhagen, I thought it might be instructive to spend some time in Zürich; a letter came back that said, "Unfortunately, I am not the Swiss Consul and cannot deny you a visa. If you come, we will just have to accept that fact." That stopped me. Later, I was told that this was an enthusiastic invitation by Pauli's standards.[7]

Few inflicted by Pauli's venom would have guessed that he also had an introspective, sensitive, mystical side. While he criticized other physical theories, he was open to speculation about numerology and the supernatural, an interest cemented through his interactions with Carl Gustav Jung.

The Secret Topography of the Mind

Jung was born in the village of Kesswil, Switzerland, on July 26, 1875. After obtaining a degree in medicine at the University of Basil, he went on to a PhD in psychiatry. The topic of his thesis, various case studies of patients with altered consciousness, including a woman obsessed with the dead, who sleepwalked to cemeteries and hallucinated about skeletons and spirits, reflected his interest in the occult. Throughout his life, he would continue to nourish a deep fascination with occult experiences.

While working under the mentorship of Eugen Bleuler—the noted psychiatrist who coined the term "schizophrenia" and founded the field of depth psychology—at the Burghölzli psychiatric hospital in Zürich, Jung soon acquired a reputation as an excellent therapist. An imposing figure who easily commanded attention, Jung pioneered a more active form of therapy, including word interplay targeted at surfacing buried issues. Having patients respond with the first word that popped into their heads, when cued with another word, has practically become a clichéd element of old-school talk therapy. Jung also coined the terms "introvert" and "extrovert" to differentiate those centered on the inner and outer lives, respectively. In short, he was a highly innovative psychotherapist.

Jung's work caught the attention of Sigmund Freud, who, in time, sought to bring him under the umbrella of the growing psychoanalytic movement. They first met in Vienna in 1907. Working closely together for six years, they found mutual interest in the importance of the unconscious mind. When in 1910 Freud founded the International Psychoanalytic Association, Jung, with his endorsement, became its first president.

By 1913, however, realizing the limits of Freud's tolerance for dissent, Jung decided to break sharply with him. Instead of focusing on the influence of childhood sexual development on unconscious motivations, Jung had turned to the notion of a "collective unconscious": cultural patterns—which he later dubbed "archetypes"—that have a common origin but resonate differently in various individuals. Such examples included fairy tales, folk myths, taboos, symbolism, religious rituals, and spiritual yearnings. These, Jung felt, in most cases impacted adult life far more than did unresolved childhood issues. To lend support to his thesis, Jung would, in time, delve into the mystical writings of alchemists, Gnostics, various Neoplatonists, Buddhist and Hindu sages, and so forth, and become an expert in mythology. He found many common elements in such mystic faiths, such as a passion for transcendental truths and the desire to unite with the divine. With the goal of exploring the relationship between individual feelings and the collective unconscious, Jung left the Vienna-based psychoanalytic movement and founded the Swiss-based analytical psychology school, connected with depth psychology, as an alternative.

The break with Freud marked the start of a period of emotional turmoil for Jung. Not only did he step down from his leadership role in the International Psychoanalytic Association, he also resigned from an academic position he held at the University of Zürich and, while remaining married to his wife, Emma Jung-Rauschenbach (whom he'd wedded in 1903), began extramarital relations with his former patient and assistant Antonia "Toni" Wolff, who became something like a second spouse for the next four decades. Moreover, he began to experience vivid, intense dreams, which fired his passion for mapping out the unconscious mind even further. He chronicled his inner struggles and fantasies during that era, through illustrations and calligraphy, in a remarkably ornate and imaginative account called *The Red Book*, first published in 2009, almost half a century after his death.

Several years before his break with Freud, Jung had memorable encounters with Einstein, who lived in Zürich at the time. Later in life, Jung would cite these as the prime motivation for an interest in unifying the realms of psychology and physics into a unified description of the mind and the flesh. It was during Einstein's early academic career—with the special theory of relativity finished and the general theory in development—that those fateful meetings took place.

Einstein's first academic position, after leaving his job at a Swiss patent office, was at the University of Zürich, beginning in October 1909. To assist with his research, he recruited Ludwig Hopf, a young physicist who had recently received his PhD under the supervision of Arnold Sommerfeld. Hopf's thesis topic was turbulent fluids, but he also had an amateur interest in the currents and eddies of human emotions. To Einstein's delight, he was also an avid pianist. Einstein loved playing violin, so they enjoyed harmonizing during breaks from research.

Hopf's psychological meanderings, including explorations of Freud's theories, led him to become well acquainted with Jung. While residing in Zürich, he took the opportunity to introduce Jung to Einstein. As a result, Jung warmly welcomed Einstein to his house—well before Einstein was famous—for dinners and discussions. At some of those meals, Bleuler was also present, as well as Adolf Keller, a Swiss Protestant theologian interested in psychoanalytic theory.

According to Jung, Einstein tried to enlighten the group about his theories of relativistic space and time. Jung found the mathematics daunting, but tried to pick up the gist of the concepts nevertheless. Once Einstein left Zürich, first to Prague and later to Berlin, they completely lost touch.

The seed was planted, though, for Jung's fascination with modern physics. Years before the concept of quantum entanglement emerged, Jung began to ponder the notion of nonlocal influences. While Einstein's theories were local, Jung latched onto the idea of relative perceptions of an objective reality.

Synchronicity Emerges

In Jung's application to psychology, our minds locally interpret an "objective psyche" composed of shared unconscious experiences that are passed on from generation to generation. Commonalities in various world religions and belief systems might be explained through such genetically disseminated primal notions. For example, veneration of mothers, fear of snakes and of the dark, and ethical concepts such as condemnation of murder, each seem to be universal experiences, which Jung attributed to that common core.

For men, in Jung's view, dreams of an anima, or prototypical female figure representing their feminine side, seemed to be prominent. In conscious life that aspect might be repressed, only to be liberated through therapy. Women often dreamt of an animus, or male archetype. Following sexist stereotypes of the times, Jung associated the anima with raw emotionality, and the animus with more refined intellectuality. He argued that a balance of the two was vital for both men and women.

Jung identified the various archetypes and imagery in alchemic texts and other occult sources, pointing, in his mind, to their universality. In dreams, fantasies, or ruminations that collective psyche might be accessed by the individual psyche, offering someone insight, hope, or fear, as the case may be. That individual side is represented as a character in dreams by the archetype of the "self."

Yet another archetype, according to Jung, is the "shadow," representing the dark side of the psyche. Embodying aspects of a person, good or bad, that he or she might be unaware of in conscious life, it might be accessible only in a heavily concealed form in dream imagery. For example, a woman might have been cruel to her sister when she was a girl, but suppressed that side of herself as an adult. In her dreams, perhaps, she might encounter a sinister figure embodying that consciously forgotten trait.

Note that Jung's concepts were pure conjecture, as he didn't offer scientific proof of such inheritance, just anecdotal evidence from case studies. Still, his speculations about the mind are fascinating from historical, philosophical, and cultural perspectives.

In 1923, Jung invited the accomplished sinologist Richard Wilhelm, who translated the *I Ching*, or "Book of Changes," into German, to deliver a talk to Zürich's Psychology Club. An ancient Chinese divination method, the *I Ching* involves casting symbolic patterns called hexagrams to prognosticate the future. Traditionally, the hexagrams are determined by collecting and sorting a set of yarrow stalks—or bamboo sticks if those are not available—in a particular way. There are sixty-four different hexagrams all in all, composed of varying arrangements of short and long horizontal lines (designated yin and yang, the two clashing elements of Taoism)—a cipher akin, in some ways, to Morse code. The array of combinations purportedly match the spectrum of possibilities in real life. From those, the option selected by chance supposedly offers insight into fate's chosen path. That selection process implies, however, that alternative worlds might be out there in which a different choice was made. The divination methods and hexagram meanings are all delineated in the book.

Wearing out his copy of Wilhelm's translation, Jung spent much time experimenting on his own. He repeatedly gathered yarrow, cast hexagrams, and tried to match them with his dreams, ruminations, and life events. He developed a keen eye for meaningful coincidence, which didn't necessarily mean such matching events were happening more often than chance would indicate. Rather, it just meant that he had a heightened sensitivity. When one is looking for coincidences they are bound to occur, due to the brain's uncanny power of pattern matching.

You can try a simple example to see how that works. Check what time it is to the nearest hour. Let's say it is six o'clock. Then focus on that number, and record how many things you come across in your daily life that match it. For instance, six windows in a certain room, or six drawers in a dresser. You'd be surprised how many coincidences arise for any hour it happens to be. Now imagine if you did that every day, you might conclude that life is full of meaningful coincidences. Jung's obsession led him to that mistaken conclusion.

As he reported in his autobiography, "During the whole of those summer holidays I was preoccupied with the question: Are the *I Ching*'s answers meaningful or not? If they are, how does the connection between the psychic and the physical sequence of events come about? Time and again I encountered amazing coincidences which seemed to suggest the idea of an acausal parallelism (a synchronicity, as I later called it)."

Jung also attempted the method with some of his patients. He noted when the *I Ching*'s predictions made therapeutic sense. For example, one of his patients had a "mother complex" and was worried that the young woman he was considering marrying might be "overwhelming" like his own mother. He asked Jung for advice on what to do. Out of curiosity, Jung cast the patient's hexagram, which reportedly read, "The maiden is powerful. One should not marry such a maiden."[8]

In tandem with becoming increasingly convinced of the power of Chinese divination, Jung embraced the sinologist Wilhelm himself in a warm friendship. Jung gladly wrote introductions for two of Wilhelm's books. Welcomed by Jung and his wife, Wilhelm often visited their stunning house in Küsnacht on Lake Zürich, where they'd engage in discussions of Taoism and other aspects of Eastern philosophy. Sadly, though, during one visit Wilhelm became afflicted with dysentery. He returned home to Frankfurt, where he was living, was hospitalized and died several months later, in March 1930.

At a memorial service to Wilhelm, held in May, Jung introduced the term synchronicity, as an acausal connecting principle. Meaningful coincidences, he emphasized, pointed to an underlying order that revealed itself in unexpected ways. In opposition to the purely causal, mechanistic universe, we must consider global influences that transcend the ordinary limitations of communication in space and time. Taoism, he felt, might teach Western minds to frame psychological interactions as not a simple interplay between people and their immediate environment, couched in terms of their past experiences, but rather transactions between the individual psyche of each and the collective unconscious of the human race, including its various archetypes and myths. Hence, he believed, dreams might be coincident with objective events—each springing up at different places at the same time. If a girl in a city was dreaming about a great conflagration, it might be the case that a boy in the countryside simultaneously had the idea to light a bale of hay on fire in a barn. There'd be no direct causal connection, but an acausal correlation through common interactions with the universal psyche. In the decades that followed, Jung would hone his ideas, eventually publishing them in his 1952 monograph, *Synchronicity: An Acausal Connecting Principle*.

Jung was not the only thinker of his day studying the meaning of coincidences. He drew in part from the work of the Austrian biologist Paul Kammerer—controversial for his advocacy of Lamarckian inheritance of environmental adaptations. Kammerer's 1919 book *The Law of Series* related dozens of improbable sequences of events gathered anecdotally from various sources. However, rather than connecting such unlikely chains of occurrences with acausal relationships due to a universal source, such as the collective unconscious, Kammerer differed from Jung in ascribing them to an emergent order in complex systems.[9] In that way, he was ahead of his time, as he anticipated a property of deterministic chaos theory, developed mainly in the 1970s and later. Also, he noted, patterns are more memorable than the lack of them. Tragically, Kammerer committed suicide in 1926, so he never had the chance to offer his own take on Jung's hypothesis.

With respect to Jung's idea of acausal connections stemming from a common background, one might immediately notice parallels with quantum entanglement—a concept coming into its own during roughly the same era. Entanglement shows how particle properties might be distantly correlated if a shared quantum state encompasses both. One crucial difference is that while Jung never proved his hypothesis to the satisfaction of the broader psychological community, quantum entanglement is widely supported by numerous carefully designed experiments targeted at eliminating any loopholes. With regard to the mind, neuroscience has shown no indication of a collective unconscious that is propagated genetically.

However, Jung's general insight that there is far more to nature than local, causal influences surely resonates in modern science, especially in tandem with his efforts to reach out to the physics community. While those interests began with his dinner discussions with Einstein, they truly took wing after he came to know Pauli.

Pauli's Crisis

By the end of 1930, Pauli was at the height of his theoretical achievements, yet an absolute emotional wreck. The preceding decade had taken him from Sommerfeld's wunderkind to one of the most accomplished and respected theorists. In particular, the exclusion principle (borne out by the discovery of the fourth quantum number, spin) and the neutrino proposal (yet to be experimentally proven, but already offering a way forward in understanding beta decay) cemented his reputation as a genius. But while the particle world was starting to shape up nicely, his own world was crashing around him.

His cascade of troubles began three years earlier, when his beloved mother committed suicide, at the age of only forty-eight, in reaction to his father's infidelity. Within a year his father remarried, wedding an artist who was in her late twenties—around the same age as Pauli at the time. Pauli derided his father's decision and nicknamed his father's new wife "the evil stepmother."[10]

By then, while Pauli's career was boosted by being appointed to a professorship at the ETH in Zürich, he had become increasingly disillusioned. For reasons not entirely understood, in May 1929 he abandoned the religion of his birth, Catholicism, by formally leaving the Church.

Pauli traveled often to Berlin, where Einstein, Schrödinger, Planck (then emeritus, but still active), and other luminaries made it one of the major hubs of theoretical physics. During one of his visits to that city he met the cabaret dancer Käthe Deppner and began dating her. She had another boyfriend at the time, a chemist, but was open to Pauli's interest. He proposed marriage to her, which for some reason she accepted, despite his being far from the man of her dreams. They wed in December 1929.

It was a troubled marriage from the start. Her interest in the chemist had not waned and she continued to see him. After a few weeks, she essentially ignored her husband. Pauli spent most of the next year in Zürich; she remained in Berlin. By November 1930 they were divorced. To Pauli's chagrin, she ended up with the chemist. "Had she taken a bullfighter, I would have understood, but such an ordinary chemist . . ."[11] he bemoaned.

With his emotional life in shambles, Pauli took up drinking and smoking heavily. He became a familiar presence at Mary's Old Timers Bar, a Zürich tavern styled after American speakeasies. It is remarkable that his neutrino idea had emerged around the same time. He was focused enough to remain productive even with his life in crisis.

Pauli's father decided to intervene, suggesting that he seek out Jung for therapy. Pauli was familiar with Jung's work, as he spoke often at the ETH. Agreeing to his father's suggestion, Pauli contacted Jung and made an appointment. By that point, he was desperate to get his inner life back on track and hoped that therapy might make a difference.

Although Pauli was expecting to be treated by the founder of analytical psychology himself, Jung assigned Pauli to his young assistant, Erna Rosenbaum, instead. Jung explained that given Pauli's issues with women he might best be analyzed at first by a female therapist. Still a student of Jung's, Rosenbaum had little experience at the time, but that was not essential. Her role was to write down Pauli's dreams until he was confident enough to jot them down himself. Her treatment of him began in

Carl Jung's spacious, custom-built house in Küsnacht, Switzerland, included rooms for psychoanalytic therapy, where he treated patients such as Wolfgang Pauli and many others; Credit: Photograph by Paul Halpern.

February 1932 and lasted about five months. Then, Pauli was placed in the driver's seat, noting his own dreams for about three months in a kind of self-analysis. Finally, Jung took over personally as his therapist for the following two years. Once Jung started treating Pauli directly, he already had more than three hundred recorded dreams to analyze, greatly aiding him in shaping his therapeutic suggestions. In addition to sharing his dreams, Pauli opened up about his emotional turmoil, erratic behavior, alcohol dependency, and issues dealing with women.

For Jung's studies of the impact of the collective unconscious on the psyche, including the roles of dreams and fantasies, he sought out subjects with vivid recall. His developing notion of synchronicity, based as it was on dynamic Einsteinian notions of space and time, could certainly benefit from a physicist's voice. Thus, analyzing a prominent quantum physicist, who happened to have complex dreams he could remember with ease, was an extraordinary find.

Realizing the possibility of a gold mine of information from a sophisticated but troubled thinker, Jung was very protective. Given his reputation as a proactive therapist who interacted far more readily with his patients than did Freud, he wanted to make sure none of his professional colleagues accused him of tampering or interference. That was another reason he assigned Pauli to Rosenbaum, and urged her not to impede or influence his dream recollections. Ultimately, either directly or indirectly, Jung compiled roughly thirteen hundred of Pauli's dreams and would make use of them (while maintaining patient confidentiality) for his research studies. Consequently, as the scholar Beverley Zabriskie wryly points out, "Readers of . . . Jung are more familiar with Wolfgang Pauli's unconscious than with his waking life and achievement."[12]

It is a mystery how Pauli remembered so many dreams in such great detail. Truly, his recall was phenomenal, and he must have trained himself in some manner. The dreams added immensely to the resources Jung could draw upon in crafting his theories. But of course it wasn't just a research project. Jung genuinely wanted to help Pauli become more aware of his stifled feelings.

The gist of Jung's treatment was to show Pauli how his emotional self, symbolized by the anima archetype, was repressed in favor of pure intellect. Pauli came to understand how his life was imbalanced. Gradually, over two years of therapy, he would become more settled—at least for a time. Finally able to maintain a mature relationship, in 1934 he married Franciska "Franca" Bertram in London. Around the same time, he decided to end his private therapy sessions. He was feeling more stable and had cut back—at least temporarily—on his drinking. Though no longer a patient, Pauli would keep up his correspondence with Jung until almost the end of his life, including sharing his dreams. They'd continue to speculate together about their significance and connections to archetypes.

With a brilliant mathematical mind engaged in unraveling the deepest questions in theoretical physics, it is not surprising that many of Pauli's dreams had geometric elements and abstract symbols. They'd often include symmetric arrangements of circles and lines, which Jung would interpret in light of his notion of archetypes. As mathematical physics fed Pauli's visions, which Jung connected with ancient symbolism, the two thinkers ended up weaving profound metaphorical connections between the two realms.

In October 1935, for example, Pauli wrote to Jung reporting a dream he had about a physics congress with many participants. Within that dream, he had thought of numerous images representing physical examples of polarization (separation of one thing into two opposites), including electric dipoles (balanced arrangements of positive and negative charges) and the splitting of atomic spectral lines within an applied magnetic field. Jung responded that his dream symbolism likely represented "the complementary relationship in a self-regulating system [including] man and woman."[13]

A saying attributed to Freud is that "sometimes a cigar is just a cigar."[14] As a physicist, wouldn't Pauli naturally sometimes have dreams with physics-related content? Couldn't it be that "sometimes a dipole is just a dipole," not a symbolic union of man and woman? Certainly. Jung, like Freud, recognized that possibility and was generally careful not to be dogmatic in his conclusions.

Another of Pauli's dreams included an ancient symbol called the Ouroboros: a serpent devouring its own tail while coiled into a circle. Such symbolism, related to the yin-yang icon of Taoism, reflects the concept of eternal destruction and rebirth, including the turning of the seasons and the recycling of the natural elements. Such a figure displays the rotational symmetry Pauli and others employed in their explorations of quantum properties. Moreover, drawing from a term in Eastern philosophy, it also forms a rudimentary kind of mandala.

Through his Eastern studies—including interpreting and writing the foreword to Wilhelm's translation of *The Secret of the Golden Flower: A Chinese Book of Life*—Jung had become well acquainted with the notion of mandalas: symmetric, geometric figures constructed from pebbles, sand, and other materials as representations of the wholeness of the cosmos. Jung correctly pointed out the universality of the concept, which spans cultures as diverse as Hinduism, Buddhism, and Jainism—with analogues in various Native American cultures. Mandalas provide sacred spaces for meditation, allowing focus of the awareness of transcendent truths, instead of the turmoil of inner monologue. As Jung described their significance:

"[Mandalas] serve to produce an inner order—which is why, when they appear in a series, they often follow chaotic, disordered states marked by conflict and anxiety. They express the idea of a safe refuge, of inner reconciliation and wholeness."[15]

One such example of mandala symbolism was Pauli's dream of a "world clock." Its imagery left a powerful impression on both Pauli and Jung. As Jung described it:

> There is a vertical and a horizontal circle, having a common centre. This is the World Clock. It is supported by the black bird.
>
> The vertical circle is a blue disc . . . divided into . . . 32 partitions. A pointer rotates upon it.
>
> The horizontal circle consists of four colours. On it stands four little men with pendulums, and round about it is laid the ring that was once dark and is now golden . . .
>
> The "clock" has three rhythms or pulses:
>
> 1. The small pulse: the pointer on the blue vertical disc advances by 1/32.
> 2. The middle pulse: One complete revolution of the pointer. At the same time the horizontal circle advances by 1/32.
> 3. The great pulse: 32 middle pulses are equal to one revolution of the golden ring.[16]

Pauli described to Jung how he found the world clock's harmony soothing. Perhaps it reminded him of the atomic mechanisms for which he played a major role in describing. For Jung, the world clock offered a unique example of a three-dimensional mandala. It helped demonstrate commonalities between an important archetype—an emblem of peaceful meditation—and the space-time concept of modern physics. It helped whet his appetite for learning more about physics even further.

Pauli kept his dream analysis quiet, for the most part. He offered permission for Jung to publish his dreams, while keeping the name of the dreamer off the record. (Once Pauli became Jung's published collaborator, the connection became obvious.) Pauli, at that point, similarly did not broadcast his growing interest in the psyche. He made an important exception for one of his closest collaborators and friends, Pascual Jordan, with whom he shared a passion for learning about the psychic world.

Parapsychology and Its Skeptics

Jordan had an extraordinary brilliance in mathematics. He was also an influential figure in the European scientific community. Though his public lecturing was hampered by a pronounced speech impediment, he was a prolific writer. His books would span numerous fields, from quantum physics to cosmology.

Politically, Jordan did not always make good choices. In 1933, he joined the Nazi Party in Germany. He later would claim that he made the choice on the basis of his career goals, rather than its ideology. He was one of the few scientists in the Party to support Einsteinian relativity, hoping to change its views on the subject from within. After World War II, during the "de-Nazification" phase in West Germany, Pauli would vouch for his character, enabling him to resume his academic work.

In 1936, Jordan wrote a primer on quantum theory, *Anschauliche Quantentheorie*. Contemporary readers might find it jolting that its final section speculates about the validity of telepathy experiments.[17] Starting in the 1930s, he had a profound interest in the emerging, and controversial, field of parapsychology (study of purported paranormal and psychic activities), founded by J. B. (Joseph Banks) Rhine, an American scientist who was a botanist by training. Banks also famously coined the term "extrasensory perception," known colloquially as ESP.

As expressed in a 1934 letter to Jung, Pauli was suspicious, at first, of Jordan's motives for embracing parapsychology—strangely blaming it on career frustration due to his speech impediment. As was often the case, Pauli was coldly judgmental:

> [Jordan] is a highly intelligent and gifted theoretical physicist, certainly one to be taken seriously. I do not know how he came to be involved in telepathy and related phenomena. However, it may

> well be that his preoccupation with psychic phenomena and the unconscious in general is due to his personal problems. These manifest themselves particularly in the symptom of a speech defect (stuttering), which almost made it impossible for him to pursue his career; this could have led to a certain fragmentation of his intellectual activities.[18]

Jordan's interest in Rhine's experiments was more solidly grounded than Zöllner's belief in Henry Slade's alleged four-dimensional feats more than half a century earlier. For one thing, Rhine was a serious scientist. Though he was interested in psychic phenomena, he took pains to try to conduct rigorous experiments (at least from his perspective) under laboratory conditions. Though critics might point to flaws in his techniques, no one alleged that he was trying to exploit public gullibility—as Slade was known to do. Thus he made the study of alleged psychic powers far more respectable. Even the skeptical Pauli eventually became intrigued by Rhine's results.

Rhine's early work, which he called the Pearce–Pratt Distance Series of ESP tests, was conducted between 1933 and 1934 at the Parapsychology Laboratory, which he and the psychologist William McDougall (who, along with Einstein, wrote one of the introductions to Sinclair's book *Mental Radio*) founded at Duke University in North Carolina. The subject, Hubert Pearce, was a Duke Divinity School student who aspired to become a minister. He believed that his mother had clairvoyant powers and was convinced he had inherited them. That possibility made him curious but also nervous. He sought out Rhine, who reassured him that his powers could be tested in a safe way under laboratory conditions.

Rhine recruited a graduate student, Joseph Gaither Pratt, to gauge Pearce's abilities to guess the symbols on specially designed "ESP cards," which he'd draw from a deck. The symbols were chosen to be distinct: a star, square, circle, cross, and two wavy lines. Pratt would select a card, which only he could see, and Pearce would remotely try to guess which of the five symbols it was. While at first they sat relatively close to each other, eventually they conducted the experiment with the two sitting in separate buildings on campus: Pratt in the physics building and Pearce in the library. By chance, Pearce should have guessed 20 percent of the cards correctly. According to Rhine, his average guessing rate was better than 30 percent. Rhine asserted that as evidence of Pearce's ESP, which he also noted by the Greek letter psi. (Curiously, and by coincidence, psi is also the symbol for the wave function in quantum mechanics.)

Psychology experiments are generally not conducted with one set of researchers and a single subject. (Rhine also tested a few more subjects around that time, but Pearce was the most promising.) Other experimenters failed to generate the same outcome, even with experiments involving Pearce. For that reason, many in the scientific community were skeptical of Rhine's results. He continued to conduct further psi experiments throughout his long career, including testing a number of other subjects.

In 1966, the British psychologist C. E. M. (Charles Edward Mark) Hansel, in his book *ESP: A Scientific Evaluation*, would call into question Rhine's experimental procedures. Hansel alleged that it would have been relatively easy for Pearce to have cheated, by quickly leaving his desk and observing aspects of the procedure through a window, for instance. In addressing Hansel's critique, Rhine, Pratt, and Pearce would each deny that any aspect of their experiments was fraudulent.

The American writer Martin Gardner, another skeptic of parapsychology, also noted that as Rhine refined his experimental conditions to address various criticisms of possible loopholes in his methods, his results became increasingly less statistically significant.

"As Rhine slowly learned how to tighten his controls," Gardner pointed out in a 1998 article, "his evidence of psi became weaker and weaker. However the evidence will not become convincing to other psychologists until an experiment is made that is repeatable by skeptics. So far, no such experiment has been made."[19]

One prominent scientist who took Rhine's experiments very seriously was the psychiatrist John R. Smythies, who worked during the 1950s at Queens Hospital in London. Known mainly for his research of the effects of hallucinogenic substances, such as mescaline, on treating schizophrenia, Smythies conjectured in a 1952 paper, "Minds and Higher Dimensions,"[20] that Rhine's experiments were evidence that consciousness was a separate substance that interacted via higher spatial dimensions. Thus, the collective unconscious was not inherited, but rather a kind of higher dimensional field, as in Kaluza–Klein theories of unification.

Pauli took note of Smythies's article. In March 1952, he wrote to Jordan:

"I . . . heard that in England wild mathematical speculations (in many dimensions) about Rhine's experiment appear in the Journal of the Society for Psychical Research. Do you know something about it?"[21]

Smythies visited Jung later that year. As he recalled, "I spent a day with Jung in Zürich in 1952. We spoke mostly about the mescaline visions and the collective unconscious. I do not recall his opinion about higher dimensions."[22]

Jung also corresponded with Rhine directly. He considered Rhine's experiments of profound importance, viewing them as possible evidence of the collective unconscious. In his mind, they showed how the divides of space and time could be breached. As Jung opined in a 1951 lecture on synchronicity:

"The Rhine experiments have demonstrated that space and time, and hence causality, are factors that can be eliminated, with the result that acausal phenomena, otherwise called miracles, appear possible."[23]

Rhine would live until 1980, defending his research against critics until the end. The field of psychology has since become far more exacting in its emphasis on controlled conditions and rigorous statistical methods.

The Nobel Citizen

The raging battles of World War II divided the physics community, with many finding themselves on opposite sides. Unlike Jordan, who joined the Nazi Party, Heisenberg, who ardently opposed fascism and anti-Semitism, tried hard to remain apolitical. Nevertheless, professing loyalty to his homeland, rather than to the party that happened to be in power, he decided to remain in Germany and serve as scientific director of its nuclear project.

Pauli was not a German citizen by birth. However, during the *Anschluss* of 1938, when Austria was annexed by Germany, his Austrian passport became invalid. Because his paternal grandparents were Jewish, it would have been calamitous for him to apply for a German passport. He tried to obtain Swiss citizenship, but was denied twice. Although, as a prominent figure, he probably still could have remained in Switzerland during the war, it would have been risky. Consequently, he elected to take up an opportunity to become a visiting researcher at the Institute for Advanced Study at Princeton, where Einstein had worked since 1933. That way, he'd be able to remain safely away from Europe during the conflict. He remained there for around five years.

The war ended in the summer of 1945. Later that year, Pauli was delighted to learn that he had been awarded the Nobel Prize in Physics. As he had begun an application for US citizenship and didn't want to leave the country, he decided not to travel to Sweden for the ceremony. Consequently, his colleagues, including Einstein, held a celebration for him in Princeton. Pauli was delighted when Einstein stood up and lauded him. In Pauli's mind, Einstein was anointing him his successor to the throne of theoretical physics. Several local physicists, including John Wheeler, also honored him with a "Nobel-Bier-Abend": an informal beer party celebrating his achievement.

A belated recognition of his Nobel would come in 1956 with an invitation to the sixth meeting of laureates at Lindau Casino in Bavaria, Germany. At that event, he'd be startled and amused by a gift from the hotel of a chocolate cockchafer, also known as a May Bug. It was the sort of odd item he'd dream about, not ingest—another curiosity in his unusual life.

In early 1946, Pauli was finally awarded US citizenship. Nonetheless, he and Franca decided to return to Zürich and resume his position at the ETH. In addition to his preference for European culture, Zürich was a far more bustling locale than quiet, leafy Princeton.

Safely back in Zürich, in late 1946 Pauli began to nurture a scholarly interest in the work of two late-Renaissance mappers of the cosmos: Johannes Kepler and Robert Fludd. Recall that Kepler and Fludd offered competing models of the solar system, with Fludd's assigning Earth and the Sun as dual central bodies around which the planets revolve—with God serving as a third center. Particularly intrigued by each of their mystical ruminations, including the numerical and geometric relationships described in Kepler's *Mysterium Cosmographicum*, including his related explanation of the Trinity in Christianity, Pauli decided to apply a kind of Jungian analysis to their writings.

As described in a letter to Jung, his initial impetus for the study was a strange dream about the Roman Inquisition's persecution of astronomers, including Bruno and Galileo. Finding himself on trial as well, Pauli desperately sends a note to Franca, passed along by a courier. Franca immediately shows up at the Inquisition courtroom and complains to him, "You forgot to say good night to me."[24] Now he's in trouble with both the Inquisition and his wife!

As the dream proceeds, a tall blond man explains to Pauli that the judges have trouble understanding the concept of rotation and revolution, and that he should explain it to them. Their ignorance about those basic principles, which Pauli understands very well, helps explain why they're so harsh in their judgments of the astronomers. The dream ends with his lamenting the astronomers' predicament to Franca and finally saying good night to her.

Pauli's interpretation, which led him to explore the differences between Kepler and Fludd, is that in the rise of modern, objective science, male thinkers ignored the anima of the world, represented by their forgotten wives. Kepler's treatment of the Trinity (in terms of the center, surface, and interior of spheres, as discussed earlier) rendered spirit as part of a mechanical system, and tried to subsume the anima into conscious experience (that is, contemplating the solar system). It thereby transferred the property of rotation from the spiritual realm (a three-dimensional mandala, like the world clock) to the material (the physical solar system). Pauli referred to that reassignment as the "objectification of rotation." That's why, in the dream, rotation was suspect.

Fludd, on the other hand, assigned separate roles for Earth, the Sun, and God, enabling a fuller picture of a balance between matter and spirit. His system didn't objectify rotation, as that wasn't a part of its physical properties. Therefore, Pauli argued, we need to take Fludd's mystical critique seriously as a counterbalance to Kepler's pure reason.

Ultimately, Pauli believed, quantum mechanics points to the need for a unified theory of the observed and the observer—of both matter and mind. In that sense, Kepler's attempts to geometrize and subsume the world of the spirit, in Pauli's view, turned out to represent a step backward, which Fludd, centuries before quantum mechanics, tried to rectify. (Note that was hardly the thrust of Kepler's work, as we saw in our previous discussion of his scientific contributions, but it made for an interesting contrast.)

Jung heartily agreed with Pauli's call for a unified explanation of consciousness and substance. He had hoped for such an explanation since his dinners with Einstein and discussions with Wilhelm. Jung's designated term for the unification of psyche and matter is *Unus mundus*, Latin for "one world."

Flood at the Institute

In 1947, Jung realized one of his own dreams—to found an institute of psychology. In establishing the C. G. Jung Institute, in Switzerland, he invited Pauli (who commanded enormous respect as a Nobel laureate) to be one of its patrons. Indebted to Jung, and eager to pursue their collaborative efforts further, Pauli was enthusiastic to participate.

Jung also invited Pauli to deliver two talks to the Psychology Club, on February 28 and March 6. Pauli was pleased to have the opportunity to discuss his theories about Kepler, Fludd, and archetypes. In the scientific community, lectures often provide the opportunity to flesh out ideas in preparation for an article. Pauli had begun to develop a long essay, "The Influence of Archetypal Ideas on Kepler's Theories," which he hoped to publish in some form.

The Jung Institute's opening ceremony, on April 24, 1948, was a gala occasion. Naturally, Pauli was one of the honored guests. Everything seemed to be in the right place for a successful event. At least at first . . .

Suddenly, there was a crash. Unexpectedly, a delicate Chinese vase, which had seemed to be secure on a shelf, had apparently decided on its own to go skydiving. It fell to the floor and shattered into pieces, releasing

The Psychology Club in Zürich, Switzerland, where Carl Jung, Wolfgang Pauli, and others gave talks; Credit: Photograph by Paul Halpern.

its watery contents throughout that part of the room. The Institute had been baptized by a flood.

Remember the "Pauli effect"? Pauli himself certainly did—destroying laboratories because of his sheer presence had become one of his trademarks. More than just a joke, he had started to take the idea very seriously.

In German, the word for "flood" is "Flut"—pronounced almost exactly the same as the name "Fludd." In English "flood" and "Fludd" are pronounced similarly as well. The coincidence spooked Pauli, who saw a meaningful coincidence between the minor deluge and his studies. Perhaps the "Pauli effect" was a genuine phenomenon after all, he started to believe, connected with Jung's theories of acausal relationships, rather than simply a subject of humor.

While putting down to paper his notions about Kepler, Fludd, and archetypes, while contemplating the odd incident at the Institute opening, Pauli encouraged Jung to refine, develop, and ultimately write up his own ideas about synchronicity. Pauli thought that concept would be of broad value.

Freeman Dyson recalled Pauli's interests and mood during that period of his life:

> I spent the summer of 1951 at the ETH in Zürich with Pauli. Since I was the only visitor, he frequently invited me to walk with him around the city after lunch, while he talked about all the subjects that interested him, physics, psychology, literature and politics. We usually stopped at a coffee-shop where he ate the ice-cream that was forbidden by his doctors. That summer he was in an unusually serene state of mind.[25]

With Nobel Prize in hand, Pauli had nothing more to prove. Still, like Einstein, he aspired to find a way to unify the world around him. His continued interactions with Jung would tie into that aspiration.

All Good Things Come in Twos and Fours

One of the remarkable aspects of the Pauli–Jung collaboration was how, by the peak of their discussions in the early 1950s, their rhetoric had begun to converge. Thanks to Pauli, Jung had become far more knowledgeable about quantum physics, including its chance aspects and the role of observers. Thanks to Jung, Pauli had become immersed in the studies of mysticism, numerology, and ancient symbolism.

Like the Pythagoreans, the duo came to be fixated on certain numbers. One of the pair's focus was the number two. Bohr's complementarity, which each embraced as a revolutionary principle, showed how nature, at its heart, included unions of opposites: waves and particles, the observer and the observed, and so forth. Curiously, Bohr was likely influenced himself by dichotomies in the works of the Danish philosopher Søren Kierkegaard (*Either-Or*) and included a yin-yang symbol in his family crest. Heisenberg's uncertainty principle embodied its own dualities—position versus momentum, and energy versus time—for which greater knowledge of one member of the pair meant lesser knowledge of the other.

During the time of his therapy, Pauli came to embrace Jung's archetypal idea of dualities. He came to accept Jung's contention that men tend to suppress their female sides (anima) and women their male sides (animus). He bought into Jung's assertion that many of his dreams had symbolism reflecting such repression.

Ultimately, those interests led Pauli to look further into such symmetries in physics, such as charge-conjugation (switching positives and negatives), parity (mirror-reflection) invariance, and time-reversal symmetries. Arguably that focus helped inspire him in 1954, in anticipation of Bohr's seventieth birthday celebration the following year, to be one of the co-proposers of the important CPT theorem (also known as the Lüders–Pauli theorem or the Schwinger–Lüders–Pauli theorem), which combines those three operations to form a single, absolute invariance.

Even more central than dualities in the numerological lexicon of Pauli and Jung are "quaternios": combinations of four. That emphasis harks back to the four elements of Empedocles, which became an important part of alchemy, and the Tetractys (triangle formed of the first four numbers) symbol of the Pythagoreans. The tetrahedron is the simplest Platonic solid. In Hebrew mysticism, such as Kabbalism, the Tetragrammaton, or sacred, unpronounceable name of God, has four letters. Mandalas typically include a central square, with symmetric features surrounding it.

In physics, the number four also makes many appearances. Recall that one of Pauli's key contributions was the set of four spin matrices (including the identity matrix)—equivalent to William Hamilton's quaternions. There are four fundamental interactions. Space-time has four dimensions. The Riemann tensor, one of the mathematical entities on which general relativity is based, has four indices, and so forth.

Along with his interest in twos and fours, Pauli shared with several other prominent members of the physics community—including Eddington, Born, and Dirac—an interest in applying mathematical reasoning to assessing why the fundamental constants have the particular values they do. The holy grail along those lines was being able to calculate Sommerfeld's fine structure constant—approximately the reciprocal of the prime number 137—from scratch. In 1929, Eddington was the first to suggest that the fundamental constants of nature, including the number of protons in the universe and the fine-structure constant, were related. He erroneously thought that the fine-structure constant was exactly 1/137, and based his calculations on that supposed fact.

As documented by the historian of science Arthur I. Miller, in 1934, Pauli mentioned in a letter to Heisenberg, and also at a lecture in Zürich, the importance of resolving the fine-structure constant quandary. Miller conjectures that "the effect of Jung's analysis opening his mind to mystical speculations"[26] might have been a factor swaying Pauli to consider that question.

Curiously, it was Pauli's student, Victor "Viki" Weisskopf, who would become privy to a mystical explanation for the number 137. Weisskopf learned from the eminent religious historian Gershom Scholem that 137 is the Gematria (Hebrew numerology) representation of the word "Kabbalah" itself.[27] Despite Pauli's interest in numerology and mysticism, he apparently did not pursue that connection himself.

An Acausal Connecting Principle

In 1950, Jung began to hone his notion of synchronicity in preparation for developing a treatise on the subject. With the help of Pauli, he hoped to shape it into a key principle acknowledged by the psychological community. As part of that goal, he aspired to develop his own emblem—a quaternio—as shorthand for how nature is connected.

Jung would schedule a two-part series of talks on the subject to be delivered at the Psychology Club on January 20 and February 3, 1951. In preparation for the lectures, on June 20, 1950, he sent Pauli a letter that included a quaternio diagram that juxtaposed causality with correspondentia (a term referring to acausal connections analogous to the Hermetic law of correspondence) and space with time.[28] It looked like:

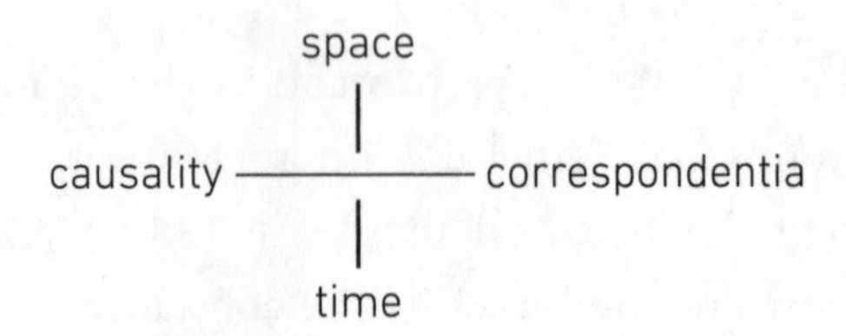

On November 24, 1950, after having a chance to think about Jung's diagram, Pauli responded with a critique about his division of space and time into opposites. Einstein's revolution, Pauli pointed out, merged space and time into a single entity—space-time—not opposites. Instead, he suggested a modified diagram (which Jung accepted with slight further modifications):

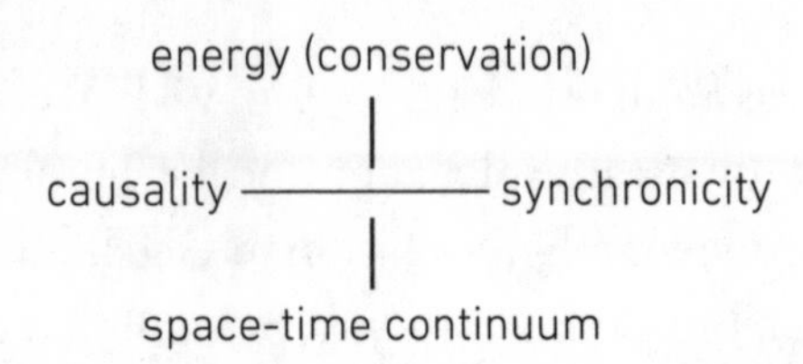

Pauli's contrast of energy (and momentum) with space-time matched the dichotomy posed by the relativistic version of Heisenberg's uncertainty principle. The more that is known about space-time, the less that is known about energy-momentum (similarly a four-dimensional entity in relativity), and the converse.

Pauli's concept of causality, called "statistical causality," which Jung came to adopt as well, was distinct from mechanistic models. He argued that given the random outcomes in certain kinds of individual quantum measurements, such as determining whether or not a radioactive sample has exhibited a single decay within a certain time frame, the law of cause and effect needed to include the notions of chance and averaging. Thus, the results of experiments are predictable only once the researchers take averages over many trials.

In the same letter, Pauli connected synchronicity with Rhine's mind-reading research:

"As you yourself say, your work stands and falls with the Rhine experiments. I, too, am of the view that the empirical work behind the experiments is well-founded."[29]

Enthusiastic about Pauli's suggestions, Jung responded with the bold proposal of generalizing synchronicity to include acausal relationships without a mental component—that is, purely physical interactions. He did not specify quantum entanglement, but surely that fell into Jung's expanded definition. Ironically, his generalization served to decouple the concept of synchronicity from the Rhine results, precisely at the same time he and Pauli were embracing them. A broad definition of synchronicity as any acausal connecting principle encourages exploration of how the universe is intertwined through symmetry and additional mechanisms other than the chain of cause and effect.

With some reservations, Pauli saw merit in Jung's expanded definition. Any extension to physical processes, he emphasized, needs to move beyond psychological terms, such as archetypes, that would not be appropriate. He wrote to Jung on December 12, 1950:

"[T]he more general question seems to be the one about the different types of holistic, acausal forms of orderedness in nature and the conditions surrounding their occurrence. This can either be spontaneous or 'induced'—i.e., the result of an experiment devised and conducted by human beings."[30]

In 1952, as the culmination of their collaboration, Jung and Pauli published a joint volume, *Naturerklärung und Psyche (The Interpretation of Nature and the Psyche)*. It included two treatises, "Synchronicity: An Acausal Connecting Principle," by Jung, and "The Influence of Archetypal Ideas on the Scientific Ideas of Kepler." Their combined work effectively outed (for anyone who read it closely) Pauli as the source of Jung's dream material. Decades later, the first part (Jung's section) would be released as a popular paperback. It contained Jung's famous anecdote of the scarab, as an example of the "meaningful coincidence" he associated with synchronicity:

> A young woman I was treating had, at a critical moment, a dream in which she was given a golden scarab. While she was telling me this

> dream, I sat with my back to the closed window. Suddenly I heard a noise behind me, like a gentle tapping. I turned round and saw a flying insect knocking against the window-pane from the outside. I opened the window and caught the creature in the air as it flew in. It was the nearest analogy to a golden scarab one finds in our latitudes, a scarabaeid beetle . . . [31]

For scientific minds, such anecdotal evidence did not help Jung make his case. Any psychotherapist who has analyzed thousands of dreams by various patients would be bound by simple chance to note coincidences at some point between elements in dreams and common events in real life, such as encounters with insects. Indeed, Jung freely acknowledged that there were alternative explanations for each of the stories he told—he just wanted his readers to note a pattern. Greater emphasis on his generalized definition of synchronicity, which encompassed physical acausal connections (such as entanglement and symmetry relationships) would have made a stronger argument for the need to move beyond pure causality. But as it stood, Jung's combination of incidents, dreams, and mythology won over few scientific adherents, aside from Pauli. One review of the book, written anonymously by an eminent mathematician, concluded: "After thoroughly studying their writings for many months now, I have come to see clearly that they are both utterly mad."[32]

End of a Dialogue

In August 1957, Pauli and Jung exchanged their final letters in their lengthy correspondence. Pauli's note to Jung that month was one of his longest, including a long exposition about symmetries in physics. He wrote it while in a state of shock. One of the symmetries of nature thought to be inalterable, parity invariance, turned out to be violated in certain weak decays. He was only happy that he didn't bet on its preservation.

While pointing to the strange fact that "God is a weak left-hander after all," referring to the evidence that certain particles twisted in only one way relative to their motion, akin to unpaired, left-handed gloves, Pauli found solace in the fact that "the 'CPT theorem' is on everyone's lips."[33] Though one symmetry in nature seemed imperfect he was confident that other symmetries would be preserved.

In the same letter, Pauli recounted a dream from 1954 in which he was in a room with a mysterious woman. The two witnessed several physics experiments involving reflections. For some reason, only they were aware that the mirror images were not actual objects. Others thought the reflections were real. He pointed to the dream as an example of his obsession with mirror imagery.

Jung responded with great interest to Pauli's letter, interpreting the dream Pauli mentioned as symbolic of the reconciliation of opposites, such as psyche and body. Jung suggests that parity symmetry violation in the weak interaction is analogous to an arbiter—called the "Third"—taking sides between two otherwise symmetrically opposite entities. For example, the Third might slightly favor psyche over body, breaking the symmetry between them. The rest of Jung's note delved into his new interest in UFOs (unidentified flying objects), which he had concluded are either real (from space) or a new kind of mythology with its own archetypes.

Why did Pauli and Jung's lengthy correspondence end with that exchange of letters, even though Pauli lived for more than a year longer? One might only speculate. A gap of months or even years sometimes happens even in the letter-writing of good friends.

Moreover, as we've seen in his attitude toward other physicists, Pauli was in his heart a skeptic. He critically examined any dogma, including Jung's. He complained to Bohr, for instance, about some of the vagaries of Jung's approach:

"The Jung school is more broad minded than Freud has been, but correspondingly less clear. Most unsatisfactory seems to me the emotional and vague use of the concept of 'Psyche' by Jung, which is not even logically self-consistent."[34]

Pauli also began to cast doubt on Rhine's methods. In a letter to Rhine, dated February 25, 1957, but received only after his death, Pauli asked Rhine about an article he had heard about that was critical of parapsychology. Rattled by the letter, Rhine complained to Jung, who tried to track down the critical piece but found no evidence of its existence.[35]

Further impetus for Pauli to disassociate from Jung was the latter's obsession with UFOs. Pauli was curious about that question, but not enough to devote the time to it for which Jung might have hoped. That period corresponded to a major collaboration on unified field theory Pauli conducted with Heisenberg. Additionally, Pauli's stamina began to decline in his final year before being diagnosed with pancreatic cancer. Thus a host of factors might have led to the end of the long, productive dialogue.[36]

8

FALSE REFLECTIONS

Navigating Nature's Imperfect House of Mirrors

It is perhaps difficult for a modern student of Physics to realize the basic taboo of the past [when] . . . it was unthinkable that anyone would question the validity of symmetries under "space inversion," "charge conjugation" and "time reversal." It would have been almost sacrilegious to do experiments to test such unholy thoughts. I remember vividly that even after I and my co-workers . . . had observed parity violation in the polarized ^{60}*Co experiments, I received a letter from Prof. W. Pauli. He believed that I would fail to observe what we designed to observe . . . [T]he great physicist Richard Feynman openly wagered that the outcome . . . would show that parity is not violated . . . As for Prof. Pauli, he was glad that he did not bet large sums of money on the possible effects.*[1]

—Chien-Shiung Wu, "Parity Violation"

AN ELEGANT THEORY, BASED ON PURE MATHEMATICAL PRINCIPLES such as symmetry, is like a precious crystal vase. An intricately crafted glass vessel might last for years, drawing attention with its symmetric beauty. Providing an ornate showcase for freshly cut flowers, it might offer a practical kind of aesthetics. Yet suppose it develops a crack near the base that allows water to seep through. Though its symmetry remains almost complete, its purpose has changed. It might well end up in a donation bin or even in the rubbish, all because of one flaw. Alternatively, there might be a way of patching it that makes it less aesthetically perfect, perhaps, but more functional. Similarly, theorists often have a choice whether to abandon a fractured structure or to find a way to make it workable.

Sometimes the crafters of a blemished work might try to deny its imperfections. Remaining attached to their original vision, they might not be able to judge what to do objectively.

For example, Schrödinger stayed committed to his original interpretation of the wave equation he developed—as a means of describing the dynamics of material pulses—even in light of evidence that quantum measurement involved additional steps. Max Born, John von Neumann, and others did him a great favor by finding a way to preserve his equation by modifying its meaning and augmenting it with ways of connecting it to realistic observations. That left Schrödinger, despite the well-deserved Nobel Prize he'd receive for the wave equation, with a bad taste in his mouth. His protests took the form of lectures and papers, such as his 1952 work, "Are There Quantum Jumps?,"[2] calling for a continuous explanation of quantum transitions instead of the discrete leaps advocated by Bohr and Heisenberg.

Pauli, similarly a Nobel laureate, would cling in his later years to the visions of nature he held dear. Hardheaded when it came to judging others' theories, he remained emotionally committed to the idea that symmetry guides the universe. In a kind of cosmic seesaw all things must balance: spin up accompanies spin down, positive charge goes hand in hand with negative charge, synchronicity offers a counterpart to causality, backward-in-time mimics forward-in-time, and mirror reflection echoes the original. In the traditions of Pythagoras, Plato, and Kepler, such was the symmetric world he cherished—a flawless, precious crystal.

After all, such symmetries, connected with conservation laws, offer the promise of fundamental, long-range connections in nature that supplement, and perhaps even transcend, its mechanistic rules. Pauli was not keen to abandon hopes for a theory of everything, based on symmetry. Indeed, even after learning that parity symmetry is broken for certain types of weak interactions, he would gravitate, at first, to a unified theory proposed by Heisenberg that took in account such developments—only to abandon the theory once other physicists pointed out its flaws.

The Faith of Madame Wu

In contrast to the dreams of symmetry in theoretical physics stood the nightmare of career asymmetry for aspiring women physicists. Throughout much of the twentieth century, the roadblocks for women to launch careers were virtually insurmountable. Even the great Marie Curie, typically the first person that comes to mind during a discussion of outstanding scientists that happened to be female, faced sexism and discrimination. In 1911, for instance, the French Academy of Sciences denied her admission, citing an all-male tradition. She was already a Nobel Laureate in Physics and about to receive her second prize in Chemistry. Maria Goeppert Mayer, only the second woman to win the Nobel Prize in Physics, similarly faced discrimination when she found difficulty getting a paid position due to her husband's already being a professor. And we saw the hurdles Emmy Noether faced in joining the Göttingen faculty.

Little girls curious about the world around us are, of course, not always cognizant of such obstacles. Rather, they might aim for the moon. In some cases, they might shatter barriers, or even break illusory symmetries.

Born on May 31, 1912, in a town near Shanghai, China, Chien-Shiung Wu (often referred to later in life as "Madame Wu") was fortunate to have a father with an egalitarian perspective. Her dad founded the Mingde School, which she attended, that was one of the few at the time to enroll girls. After completing her schooling, she went to the prestigious Nanjing University—where one of Curie's former students happened to teach—and received a physics degree.

In 1936, Wu set sail for California, where she enrolled at Berkeley. Upon completing her PhD in 1940, she was unable at first to find a professorship. She accepted several different instructorships (at a lower level than her qualifications) and participated in the Manhattan Project before finally obtaining a professorship at Columbia. There she established a reputation as an expert in beta decay, helping verify Fermi's theory of how it transpires.

Experimentalist Chien-Shiung Wu (right) and theorist Wolfgang Pauli. Wu showed that parity symmetry is not conserved in some weak interaction decays; Credit: AIP Emilio Segrè Visual Archives.

In the early 1950s, among the new particles discovered in cosmic ray debris were two types of mesons (particular kinds of hadrons that have integer values of spin) that seemed similar in every way except for the manner in which they decayed. Dubbed the "tau" and the "theta," the former decayed into three pions (another type of meson) and the latter into two pions. Each decay mode conserved total charge, and every other characteristic, except for spatial profile.

In 1956, two young physicists, Tsung-Dao Lee and Chen Ning "Frank" Yang, coauthored a revolutionary paper, "Question of Parity Conservation in Weak Interactions," speculating that both the "tau" and the "theta" were precisely the same particle—which is now known as the "K meson," or "kaon." What was different about the two modes of behavior was simply that they weren't carbon copies or reflections of each other. They speculated that parity—mirror reflection—was not conserved in certain weak interactions, even though it was clearly conserved for electromagnetism and the strong interaction. They proposed decay experiments to test their hypothesis.

In developing their paper, Lee, who worked at Columbia, consulted with Wu. She immediately became fascinated by the idea of testing parity violation. Brainstorming with Lee, she suggested the idea of polarizing (separating the spin modes) of a Cobalt-60 radioactive source. Wu worked closely with researchers at the National Bureau of Standards (NBS) in Washington, DC, an ideal place to carry out such high-precision measurements.

Once Lee and Yang's article appeared, the clock was ticking on testing their hypothesis. Wu and her husband were planning a trip to China for the Christmas holidays, and had already booked two tickets on the Queen Elizabeth ocean liner for the long voyage. Convinced of the importance of Lee and Yang's result and worried that another team might beat her to confirming it, she canceled her own plans, sending her husband off on his own, and headed down by train to the NBS instead.[3]

To polarize the Cobalt-60, a sample needed to be cooled down to frigid temperatures close to absolute zero and then placed in a strong magnetic field. The spin states of the nuclei (a superposition of the states of the neutrons and protons within them) would generally align along the direction of the field, minimizing their overall energy. Wu and her team then compared how many of the electrons released by the nuclei through radioactive beta decay headed toward either the upward or downward directions. They noted a significant discrepancy between those two counts, showing that nature favored one mode of decay over its mirror image.

The Cobalt-60 was acting like a lawn sprinkler that for some reason soaked one side of a field while only moistening the other—or an ambidextrous baseball pitcher that favored left-handed throws nonetheless. Despite the sample's overall symmetry, its spray of electrons was decidedly asymmetric. Lee and Yang were right. Parity was violated in certain types of weak interactions.

Wu returned to Columbia and shared the news with Lee, who excitedly conveyed it to Yang. Lee also brought it up at a customary weekly Chinese lunch, in which other members of the physics department were present. Among them was Leon Lederman, who immediately thought of a way of testing parity violation using muons (a more massive cousin of electrons) produced in a cyclotron. He conducted such an experiment along with his older colleague Richard Garwin and his graduate student Marcel Weinrich. That second team similarly found parity violation. To be fair to Wu's team, she submitted her paper first and the second team followed, giving her priority for such an important discovery. Nonetheless, the Nobel Prize for the discovery of parity violation would go to the theorists alone, Lee and Yang. Wu's non-inclusion continues to be cited as an example of bias against women scientists. Nevertheless, she would receive numerous high honors throughout her lifetime, including the National Medal of Science and election to the American Physical Society as its first woman president.

Southpaw Neutrinos

Shortly after the teams' discoveries, Pauli learned about the results from Georges M. Temmer, a colleague of Wu's at the NBS, who happened to be visiting Europe at the time. After Temmer informed Pauli that parity symmetry was no longer conserved, he remarked, "That is total nonsense!" Temmer responded, "I assure you the experiment says it is not." Pauli then insisted, "Then it must be repeated."[4]

It was a deep wound to Pauli's sense of natural order that a symmetry as basic as mirror reflection was no longer intact. Yet, like his hero Kepler, when confronted with the non-circularity of the planets' motions, he came to accept the verdict of the scientific method. As numerous experiments verified parity violation in the weak interaction, even Pauli became convinced.

Ironically, it was the particle most associated with Pauli, the neutrino, that turned out to be the leading culprit. Related to parity is the concept of chirality, or handedness. Neutrinos, as it turns out, are always left-handed, or "southpaws" in colloquial speech, meaning that their spin states and direction of motion are anti-correlated. Conversely, antineutrinos are always right-handed. Therefore, in the various types of beta decay, and other forms of weak decay, when neutrinos or antineutrinos are produced, the lack of their opposite handedness in nature breaks the mirror reflection symmetry.

It would be like two identical, left-handed twins standing on opposite sides of an empty frame pretending to be mirror images of each other. One of them insists on putting a baseball glove on her left hand. She asks her sister to find a right-handed glove to put on her right hand to preserve the illusion of being a mirror image. But they don't have any. The sister puts on a left-handed glove too, and the parity symmetry is broken. Like interactions with neutrinos, they can't quite comprise perfect mirror images.

Electrons, on the other hand, and most other elementary particles, come in both right-handed and left-handed varieties. It is a true mystery, not yet resolved, why neutrinos ended up as only southpaws.

Even with parity symmetry violated in certain weak decays, many physicists remained confident at the time that a related symmetry, time-reversal, remained intact. As fate would have it, that supposition would similarly be proven wrong, through kaon decay experiments conducted by Val Fitch and James Cronin in 1964.

The notion of time-reversal symmetry has a venerable history, dating back to classical physics. Newton's laws of motion are perfectly symmetric in time. That means that the microscopic interactions of classical particles, if filmed, would appear the same if the movie were run forward or backward in time. By the nineteenth century, the law of non-decreasing entropy showed that on a macroscopic level, the buildup of waste energy coincides with a distinct arrow of time.

The fundamental equations of quantum mechanics, pertaining to microscopic objects, are similarly time-reversal invariant. The process of measurement, however, triggers wave-function collapse, which has a preferred direction in time. One might argue, though, that it is the interaction with macroscopic observers that generates that arrow. Therefore, one might conjecture that nature, on its smallest scale, is indifferent to clock direction, and that the macroscopic arrow is a product of how we observe it.

In the early 1950s several theorists, including Pauli, Julian Schwinger, Gerhart Lüders, and John Stewart Bell, independently proposed CPT (charge-conjugation, parity, and time-reversal), symmetry invariance. Charge-conjugation means flipping from positive to negative, or the converse. It leaves electrically neutral objects the same. CPT invariance involves performing three operations in succession to return to the original: charge-transformation, along with a parity (mirror-image) transformation, and a reversal of the direction of time. As we've seen in the case of a positron heading toward an interaction site, enacting all three transformations always leads to an equivalent depiction. Flip the positron's charge to negative, send it in the opposite direction, and backward in time and one gets a time-reversed electron slinking away from the interaction—which, as explained, is equivalent to the original positron.

CPT invariance, to date, remains a robust symmetry of nature. Its components, however, have been shown to have exceptions. The invariance of CPT implies that if CP is broken, T must also be broken. The Fitch–Cronin experiments of 1964 would show that if CP is violated, therefore T must be violated as well. Ah, the ephemeral nature of perfect symmetry, from melting snowflakes and wilted sunflowers to time-reversal discrepancies.

Superpowers

We don't have to look to the microscopic world to find curious quantum effects. Around the same time in the mid-1950s, when various researchers were exploring the nuances of quantum symmetries, considerable progress was made unraveling key mysteries about unusual types of substances—superconductors and superfluids—examples of a phenomenon called quantum coherence.

Superconductors are materials that completely lose their electrical resistance below a certain temperature. In normal resistors, such as the filament of an old-fashioned light bulb, electrical flow is hindered by the pinball-machine-like bouncing of electrons off intervening atoms of the material. The electrons lose energy, which in the case of a filament, leads it to glow. However, in the case of superconductivity, electricity slinks right through as if there were no obstacles at all. That's because electrons find it energetically favorable to form couples, called Cooper pairs, that effectively act like bosons. At the ultra-cold temperatures of ordinary superconductors, these pairs are protected by an energy gap from splitting apart—meaning that there is not enough thermal energy available per pair to split them apart. Unlike fermions, bosons have the capability of sharing the same quantum state. Consequently, quantum coherence enables such paired electrons to proceed in tandem, essentially oblivious to their environment, like seasoned soldiers effortlessly marching in lockstep through a hazardous zone.

Discovered experimentally in 1911 by the Dutch physicist Heike Kamerlingh Onnes, it took almost half a century for the full theory of superconductivity to emerge. Leon Cooper's concept of paired electrons fit well with related work by John Bardeen and J. Robert Schrieffer, leading to a groundbreaking joint paper. Published in 1957, the "BCS theory" would be honored in 1972 with its three authors receiving a shared Nobel Prize in Physics.

Strongly related to superconductivity is the similarly low-temperature phenomenon of superfluidity: zero viscosity, meaning perfect flow. Helium-4, and the rarer isotope Helium-3, when liquefied, each display that characteristic. Vortices formed in such superfluids by stirring them persist indefinitely, unhindered by conventional resistance. Such behavior is similarly caused by the clumping of effective bosons into a shared quantum state.

Remarkably, such cohesive states in some cases can be indefinitely large, and can occur, under the right circumstances, at high temperatures as well as low. For example, in recent years astronomers have speculated that the interiors of neutron stars, the highly dense, collapsed cores of massive stars made of compressed nuclei, resemble superfluids despite temperatures in the hundreds of millions of degrees. In that case, their cohesive states would be miles across.

Quantum coherence, in superconductivity, superfluids, and related phenomena, shows how synchronous correlation—what Pauli and Jung included under the umbrella of synchronicity—is as much a part of nature as causal links. Sometimes nature likes to coordinate using shared quantum states instead of signals. The spread of light in the cosmos is only part of its web of connections.

Uniformity and Its Discontents

The transformations between superconductors, superfluids, and their ordinary equivalents each involve phase transitions—a sudden switch from one form of order, with its own symmetries, to another, with fewer symmetries. Each phase transition takes place at a certain critical temperature, representing the point when the average thermal energy per particle (or pair) is enough to break the original symmetries. One doesn't need to look far to find a more mundane example of phase transitions in nature. Ice cubes, once heated to their melting points, cede their crystalline symmetries and become puddles of water. If on a hot, humid day, the temperature suddenly chills, one might be drenched in a downpour.

Beginning in the 1950s and especially in the 1960s, the search for an ultimate theory of the natural forces began to turn to the notion of symmetry-breaking in field theory. Theorists began to explore the idea that the hot early universe was far more symmetric on the fundamental level than the state of the cosmos today. But instead of treating the broken symmetries as a bad thing, they began to embrace it and explore its possibilities for explaining how ultimate unification evolved into the complex world of particles and interactions we see today.

For example, in the very early universe, the number of matter and antimatter particles may well have been balanced. Some type of symmetry-breaking may have disrupted that equilibrium. Hence, the bulk of the bodies we observe in nature are made of antimatter, not matter.

In another case of symmetry-breaking, the unified theory of the natural interactions may well have involved all massless fields. Then, as the universe cooled down, some of the fields gained rest mass, while others, such as photons, remained massless. The specifics of how that happened were developed in the early to mid-1960s by several teams of researchers and would become known as the Higgs mechanism—named for one of those theorists, Peter Higgs. In that procedure, a transformation of the vacuum state of the universe induces an energy field, called the Higgs field, to spontaneously break its original symmetry. In doing so, it couples with a number of fundamental particles, causing them to acquire mass.

Hence while uniformity and smoothness are psychologically appealing, it is the unevenness spawned by symmetry-breaking that has led to critical developments in nature. Absolute symmetry, if sustained forever, could not have produced the conditions for life, such as the segregation of the natural forces into their various strengths and the dominance of matter over antimatter. We owe our very existence to shattered regularities.

Perhaps it is not a coincidence that considerable progress in unification was made in the years following the death of Einstein in April 1955. Before then, unification was tainted by his quirks and shortcomings. His quixotic drive for unifying two of the natural forces, while essentially ignoring the nuclear interactions, quantum processes, the growing zoo of elementary particles, and so forth, had inspired raised eyebrows and bemused expressions, but precious little confidence. Only a few other physicists, notably Oskar Klein, Pauli, and Schrödinger, dabbled in unification ideas, and were inevitably compared to Einstein. Therefore, in some ways, Einstein's passing created an opening for novel ideas.

By 1957, Heisenberg certainly wasn't a fresh face in physics. He was still primarily known for the uncertainty principle, some thirty years earlier. In the interim, along with getting the Nobel Prize, perhaps his best-known contribution to science was a negative one: heading the German nuclear effort under Hitler. While he insisted after the war that he had dragged his feet and made no efforts toward a bomb, a segment of the international physics community suspected that there was more to the story and essentially shunned him.

Yet, from his perspective, the criticism was unfair. He had never been a zealot, Nazi Party member, or anti-Semite. On the contrary, some of the most doctrinaire Nazi scientists, such as Philipp Lenard, had criticized him during the Nazi era for his continued support of Einstein. After the war, he had become an avid pacifist and strong supporter of international cooperation, including advocacy of the founding of CERN (European Organization for Nuclear Research), established in 1954. His involvement took him on visits to Geneva, where it was headquartered. While in Switzerland he would sometimes stop by Zürich and consult with Pauli.

In his later years, Heisenberg had become increasingly philosophical in his writings. Drawing on the readings of his youth, he began to couch his own exploration of the fundamental nature of reality within the context of the ancient Greeks, particularly Aristotle and Plato. By moving beyond realism and causality, while emphasizing principles such as symmetry, he began to see the quantum program as a natural extension of Plato's notions of the realm of forms and the perfection of symmetric geometries.

As the historian of science David Cassidy noted:

> It was during the postwar years, as West Germany attempted to overcome the chaos of defeat, that Heisenberg made several claims that he had read and deeply discussed Plato in his early years. And he argued that Plato's idealism and Aristotle's "substance" could have direct implications for postwar elementary particle physics. It was also in that era that he came closest to linking his physics with ancient philosophy, but it was only a tenuous link at best—more for public consumption than for private theorizing.[5]

Heisenberg contrasted the abstract conceptions of Aristotle and Plato with the classic atomism of Democritus, which he dismissed as the seed of the materialist, mechanistic tradition that quantum physics had transcended. He thus joined a venerable tradition of scientists who dabbled in Neoplatonism in its many guises.

"The particles of modern physics," Heisenberg wrote in one of his final works, "are representations of symmetry groups and to that extent they represent the symmetrical bodies of Plato's philosophy."[6]

A more recent "philosopher" of sorts Heisenberg hoped to supersede was Einstein. Heisenberg always admired Einstein's brilliant mind, while seeing him as misguided in his dismissal of quantum uncertainty. Still, Einstein's notion of uniting the laws of nature was appealing. Heisenberg began to think of unification in a way that extended quantum field theory, rather than reinventing the wheel.

One of Einstein's major goals was to express all known particles and interactions in terms of pure geometry: as ripples in the fabric of space-time. Heisenberg homed in on the concept of a universal energy field. Each particle or interaction would manifest itself as a property of that field that emerged through measurement. Independent particles and fields wouldn't really exist; they'd each be aspects of the same universal field. Scattering data from various interactions might be used to probe the symmetries and other properties of that field, using a mathematical spreadsheet called an "S-matrix" or scattering matrix. Hence, in the quantum tradition, what we observe would not be the fundamental reality, but only an indirect reflection.

Many quantum field theories are plagued with infinite terms that make no sense physically. The standard procedure for eliminating such maladies is the cancelation of terms known as renormalization. As Heisenberg developed his own field theory that turned out to be nonlinear (meaning the equations contain squared terms and other higher powers, making it so the sum of two solutions might not itself be a solution), he could not use the same methods. Instead, he artificially set a smallest-length scale that prevented the possibility of division by zero, thus blocking infinities.

Heisenberg decided to consult with Pauli several times about his ideas. At first they disagreed on some aspects of the symmetry properties of the energy field. Eventually, though, they found common ground.

By late autumn of 1957, Heisenberg was starting to feel more confident. During a visit with Pauli in Zürich, he presented a nonlinear field theory that incorporated multiple symmetries, including spin, isospin, and Lorentz (the symmetry of special relativity), while at the same time including the possibility of parity symmetry violation, in line with the work of Lee, Yang, and Wu. The particles we see would condense as especially robust, stationary states of that field like chunks of ice that arise as a lake freezes. The natural interactions would emerge via the probabilities of certain types of transitions taking place.

Surprisingly, given his typical skepticism, Pauli found some merit in the approach and offered his help to his old friend to flesh out the idea mathematically. Together they strived to work out its details. As they toiled in tandem, Pauli's confidence grew that they might have found the universal equation that had evaded Einstein's grasp.

The Artifice of Interaction

General relativity and quantum mechanics emerged within a decade of each other, from the mid-1910s until the mid-1920s. In the former, space-time is the main arena, which, in its curvature, physically steers the paths of light rays and particles. In the latter, especially according to the worldview promoted by Heisenberg, abstract Hilbert space is the crucial playing ground. Practical-minded scientists—such as laboratory researchers—recognize both venues and appreciate the importance of each to the understanding of physics. Hence, one important area of theoretical physics describes quantum field theory in curved space-time—in essence combining general relativity and quantum mechanics by paying homage to the rules of both. Combining both theories without compromising either one follows in the spirit of Bohr, who argued strongly for the special importance of the classical world of the observer (which follows relativity under extreme conditions such as high velocities and gravitational fields) in its role of taking measurements of the atomic realm.

Grand unifiers, however, such as Einstein, Heisenberg, and Pauli, sought a comprehensive theory of everything that seamlessly accounted for the atomic, mundane, and cosmic realms through simple principles. While Einstein tried to mimic the quantum rules using a non-quantum approach (extensions of classical general relativity), Heisenberg set out for himself a markedly different task: to begin with an energy field in Hilbert space that obeyed certain symmetry groups, set certain constraints, and reproduced the various types of interactions in nature.

In Einstein's world, the universe is a frozen block of space-time. At each point, however, one might trace a light-cone that sets the limits of causal connections with other points. Locally, at least, no communication might move faster than light. (Exceptions arise only globally, in cases, such as wormholes, where there are twists and tears in the fabric of space-time.) Hence, causality is built into the structure of reality.

By relying on connections mediated by an energy field in Hilbert space, Heisenberg and Pauli began with a kind of synchronous existence. Clearly, it was not the synchronicity described by Jung's notion of a collective unconscious, as Heisenberg was not a Jungian. But it offered a kind of eternal background reminiscent of Plato's realm of forms. Causality thereby needed to be superimposed, by making faster-than-light interactions extremely unlikely. The distinction is a bit like the difference between driving below a certain speed because of rough terrain versus doing so because of imposed traffic rules. Other forms of quantum field theory imposed such constraints, so that part was not unusual.

What was indeed distinct about the Heisenberg–Pauli model was the notion of a single fermionic field from which all matter, energy, and natural interactions would coalesce as special cases. A particle such as the electron, proton, or neutrino would have no independent existence, but rather arise only when conditions in the universal field favored its production and persistence. Electromagnetic and other interactions emerged via high-probability correlations between such ersatz particles. Hence light would be a phantom wrought by interactions among elements of the universal field, transpiring with a time delay most likely consistent with matching the known speed of light. Following Noether's theorem, symmetries of the field would guarantee that all known conservation laws were obeyed. Lack of parity symmetry would be imposed in special cases. Thus, the "real world" would prove to be the ultimate illusion.

Heisenberg was delighted to channel Plato in offering an update on the realm of forms. As for Pauli, the theory's emphasis on symmetry, universality, and synchronous connections gave him metaphysical shivers of ecstasy. As a boon, it purported to offer a way of justifying the Sommerfeld fine structure constant, with its sacred inverse of the number 137—a goal long sought by Pauli and others.

Crazy for Unity

In January 1958, physicists from around the world, including Niels Bohr and many other notables, flocked to New York for the annual American Physical Society (APS) convention. Pauli and his wife, Franca, who were already planning a stay in the United States, sent a message by cable to the physics department at Columbia University explaining that he wished to deliver a talk about his research with Heisenberg. Sensing, perhaps, an opportunity to engage with Pauli about the new horizons in physics opened up by the discovery of parity violation, Wu hurriedly invited him to speak and organized the event.

Because of the APS meeting, only a short subway or taxi ride away, Pauli's seminar was packed. Bohr was one of many who jammed into the room. As attendees recall, as Pauli delivered the lecture, he began to seem exceptionally unsure of himself. He didn't seem to believe what he was saying. Somehow he hadn't thought it through. Stanley Deser, who was in the audience, recalled:

"The final episode in [Pauli's] tragically short life came in 1958 when he had erroneously embraced Heisenberg's crazy fermion model and talked about it at an overflow session of APS, and slowly realized it was nonsense. None of us spectators will ever forget it!"[7]

The physicists Freeman Dyson and Jeremy Bernstein were also in attendance. At one point, according to Bernstein, Dyson leaned over and said to him, "It is like watching the death of a noble animal."[8]

At the end of the talk, Pauli reportedly told the audience that the theory might well be crazy, alluding perhaps to the initial reaction to some of his earlier ideas, such as neutrinos (which had just been discovered). That apparently prompted Bohr's response: "We are all agreed that your theory is crazy. The question that divides us is whether it is crazy enough to have a chance of being correct."[9]

By all accounts, the seminar would prove a sad denouement to Pauli's professional career. Pauli's enthusiasm for the theory started to chill. Despite its glistening symmetry, it wasn't very predictive. Without imposing extensive, artificial constraints, it didn't reproduce the known features of the subatomic realm. The Higgs mechanism, a natural way of generating mass, had not yet been invented. Instead it offered little explanation as to why some particles are massive and others massless.

Pauli surely wondered how he had not seen its flaws. In his mind, it was certainly not ready to be published. Nevertheless, he had agreed that Heisenberg could issue a preprint (unpublished version) of the work, as long as it was clearly not the final product.

Heisenberg, back in Germany, remained excited about the project. To Pauli's dismay, his eagerness to spread the word got the better of him. In February, shortly before he sent out copies of the preprint to leading physicists, he gave a talk in Göttingen about the theory. A reporter, who happened to be in the audience, got overly enthused. Newspaper accounts about the "world formula" soon appeared, with one noting that "Professor Heisenberg and his assistant, W. Pauli, have discovered the basic equation of the cosmos."[10]

Pauli, who had once mentored Heisenberg, was mortified to be described as his "assistant." He became even more incensed when he heard Heisenberg deliver a radio address about the theory, asserting that only the "technical details" were missing. Pauli was aghast to have his joint work misrepresented as being nearly complete, when it was only rudimentary.

Pauli wrote to his student Charles Enz, "Have you . . . heard of Heisenberg's radio- and newspaper-advertisement, with him in the principal role of super-Einstein, super-Faust, and super-human? His passion for publicity seems insatiable."[11]

To gain a kind of revenge, Pauli took out a drawing pad, sketched a rectangle to look like an empty picture frame, and wrote under it the caption "This is to show the world that I can paint like Titian. Only technical details are missing." He mimeographed the sketch and sent copies to a number of leading physicists, including George Gamow.[12] More formally, he also sent a letter to all of the physicists on Heisenberg's preprint list withdrawing his support for the theory.

To make relations between the two collaborators even tenser, in July, Pauli was chair of a session at a CERN conference, for which Heisenberg was one of the speakers. Before Heisenberg spoke, Pauli offered a scathing introduction: "This session is called 'fundamental ideas' in field theory, but . . . what you shall hear are substitutes for fundamental ideas, and it works in the same way as I am the substitute for a rapporteur. So, you will also see there are two kinds of ignorance; the rigorous ignorance and the more clumsy ignorance."[13]

The rift cast a dark shroud over their friendship when, only a few months later, as fate would have it, Pauli suffered an untimely death due to cancer. Heisenberg, who was very sensitive, would remain bitter about Pauli's criticisms for quite some time afterward.

Pauli would not live to see the end of 1958. On Friday, December 5, as he was delivering a weekly seminar, he began to feel sharp pains in his stomach. After heading home and trying to rest, these persisted throughout the night and into the morning. He checked into the Red Cross Hospital in Zürich to be evaluated and was assigned a room.

Primed by his experiences with Jung to be keenly observant of numerological signs, he noted the room number—137—the approximate inverse of Sommerfeld's fine structure constant.[14] Viewing that, perhaps, as an omen of a failed quest about to reach its end, he did not take that well. To this day, despite many attempts, no one has found a reason that the fine structure constant is so close to the inverse of a fairly commonplace whole number.

Diagnosed with late-stage pancreatic cancer, Pauli's prognosis was poor. He was understandably very distressed. Nevertheless, his biting humor emerged when he told Enz, who was visiting him, that he expected Heisenberg would soon inform Gamal Abdel Nasser, the powerful new pan-Arabist president of Egypt (who was much in the news for uniting that country with Syria) about his unified theory. Pauli also reportedly contacted Jung's secretary, hoping for a meeting with the psychoanalyst who played such an important role in his life. That final encounter was not to happen.

On Saturday, December 13, surgeons attempted to remove the large tumor that had engulfed his pancreas, but ultimately concluded that it was inoperable. He died two days later, on the morning of December 15, and was cremated on December 20. The physics community was stunned by the sudden loss of one of its leading lights. Heisenberg was still so upset by Pauli's behavior earlier in the year that he skipped his old friend's funeral.

As Pauli was only middle-aged when he died, many of the important people in his life survived him. On June 6, 1961, less than three years after Pauli's death, Carl Jung died of heart failure in his Swiss home near Zürich at the age of eighty-five. Heisenberg succumbed to cancer of the kidneys and gallbladder on February 1, 1976. Franca survived her late husband by almost three decades, aiding historians immensely by helping assemble his papers, and died in July 1987.

Pauli's legacy has mainly been centered in the physics community, where he is remembered for the exclusion principle, the neutrino, and other pivotal contributions. Surprisingly, he has remained little known by the general public. Having witnessed how the theories of Heisenberg, Schrödinger, Bohr, and especially Einstein were heralded in the press in an overblown fashion, perhaps it was Pauli's caution that largely kept him out of the news. He always preferred the genuine recognition of his colleagues over a misguided portrait in the press. And, despite his cantankerous nature, such respect he certainly had, and still has to this day.

9

REALITY'S RODEO

Wrangling with Entanglement, Taming Quantum Jumps, and Harnessing Wormholes

There is nothing mysterious about quantum entanglement except for the fog created by philosophers trying to interpret it.[1]

—Freeman Dyson (remark to the author)

It is rather unfortunate that Wolfgang Pauli didn't live several decades longer, as he would likely have been impressed by the scientific progress made. The era that followed his untimely passing—particularly the 1960s through the 1980s—represented some of the greatest breakthroughs in particle physics, quantum field theory, quantum measurement theory, and related fields. Some of the key ideas he pioneered, such as the role of symmetry in physics, and (in tandem with Jung) the need for a comprehensive description of acausal connections, moved to center stage in physical discourse—the latter in a focused effort to test the limits and possibilities of quantum entanglement.

Pauli had long dreamed of a unified field theory guided by symmetry considerations. While a full, satisfactory explanation of all four of the natural interactions—electromagnetism, the weak force, the strong force, and gravitation—has yet to be developed, the first two have been fully unified into electroweak theory. In tandem with the strong force, which has been described through similar language, the combination is called the Standard Model. It is brimming with symmetry, but cleverly takes into account the violation of parity and CP symmetry in certain types of weak interactions, through built-in imbalances of its terms. A quantum theory of gravitation that is broadly accepted by the community remains elusive. Therefore, gravity continues to remain the odd theory out.

Early attempts to model the strong force identified the notion of vibrating, energetic strands called "strings" as a means of modeling the bosons conveying that interaction between pairs of hadronic (strong-force-feeling) fermions. While quantum chromodynamics, developed by Murray Gell-Mann and others, ultimately replaced the bosonic string idea, in the early 1970s the physicists Pierre Ramond, André Neveu, and John Schwarz showed how to define fermionic strings and unite them with bosonic strings in a proposed new kind of symmetry, called supersymmetry. It made it possible to describe fermions, which obey Pauli's exclusion principle, and bosons, which do not, by means of a common mathematical construct. Ultimately that would inspire supergravity, superstring theory, and M-theory—attempts to incorporate gravitation, along with the other forces of nature, in a unified quantum field theory without the infinite terms that plagued earlier attempts. Although experiments at the Large Hadron Collider (LHC) and elsewhere have yet to provide a hint of evidence for supersymmetry, many theorists remain optimistic, largely because of the concept's mathematical elegance.

John Bell's Quantum Litmus Test

Concurrent with the extraordinary progress made in particle physics and field theory since the death of Pauli, there have been brilliant gains in the understanding of quantum processes in general. One of the greatest theoretical leaps, made in 1964, was the physicist John Stewart Bell's mathematical litmus test to gauge the nonlocality of entanglement by distinguishing hidden variable theories, such as the De Broglie–Bohm pilot wave formulation, from standard quantum predictions. It enabled, for the very first time, the possibility of constructing experiments that would

Quantum theorist John Stewart Bell (1928–1990) and his wife, Mary Bell, in Amherst, Massachusetts, summer 1990; Credit: Photograph by Kurt Gottfried. Courtesy of the AIP Emilio Segrè Visual Archives, with permission of Kurt Gottfried.

distinguish quantum models with local realism—Einstein's preference—from those that revealed certain properties remotely and upon measurement—Bohr's position.

Without rigorous experimental testing, robust scientific hypotheses and superficial speculations often cannot be distinguished. As we've seen in numerous examples, from Pythagorean claims about an unseen "counter-Earth" to Tait's visions of souls composed of ether, and from Slade's alleged four-dimensional manipulations to Sinclair's assertions of his wife's mental feats, a lack of scientific rigor can lead to nonsensical beliefs. Even the Rhine experiments were not reproducible by other labs, including those critical of his work, and thereby were suspect. Consequently, accepting quantum entanglement as scientific just because it was proposed by scientists, but without an ironclad way of gauging it, was taking a big risk. Luckily, Bell offered a means of assessing its reality.

Bell was born in Belfast, Northern Ireland, on July 28, 1928. From an early age, he was fascinated by the workings of science. When only eleven, he excelled in the high school entrance exam and became a pupil at Belfast Technical High School. Upon graduation, he worked a year as a technician and then continued as a student, at Queen's University, Belfast. There he began to learn about the oddities of quantum physics, including Heisenberg's uncertainty principle, which he thought was poorly formulated. In Bell's mind, the way his instructor, the physicist Robert Sloane, presented the theorem, it sounded like one could arbitrarily set the ranges of position and momentum, as long as they obeyed Heisenberg's formula. Bell believed there needed to be a more technical explanation, based on the lab apparatus. When Sloane began to justify Heisenberg's formulation, Bell argued with him vehemently, accusing him of "dishonesty."

"I hesitated to think [the theory] might be wrong, but I knew it was rotten," recalled Bell in an interview with Jeremy Bernstein. "That is to say, one has to find some decent way of expressing whatever truth there is in it."[2]

For practical reasons, Bell's aspirations to revamp quantum theory would need to wait for a time. Upon receiving his undergraduate degree, in 1949, he decided to work for the atomic energy establishment in the UK, centered at Harwell. There he met his life partner Mary Ross, a bright and accomplished fellow scientist who would collaborate with him on various projects. They would marry in 1954.

Taking a leave from Harwell, in 1951 Bell began to conduct theoretical research at the University of Birmingham, a project that would ultimately lead to a PhD in 1956. While completing his doctoral research, he independently proposed the CPT theorem, but Lüders and Pauli each published first and gained the bulk of the recognition. In 1960, he and Mary joined the research faculty at CERN, where he would rack up his major accomplishments.

As the physicist Kurt Gottfried, who worked with Bell at CERN for many years and knew him very well, recalled:

"He was a very inquisitive mind and he was a very deep thinker."[3]

In his theoretical research, Bell belonged to the small minority of physicists who were sharply critical of the orthodox view of quantum mechanics, as presented by Bohr, Heisenberg, Born, John von Neumann, and others. He thought Einstein's critiques were apt and that the quantum world needed a more comprehensive description.

When in 1952 David Bohm expressed the EPR experiment in terms of spin states and developed his "pilot wave" theory (also attributed to De Broglie) that included hidden variables, Bell was elated. Finally, he believed, there was a way of showing the underlying mechanisms for how entanglement truly worked. Determinism and realism in nature, he thought at the time, could finally be restored.

One mystery about Bohm's formalism was why an earlier theorem, proposed by von Neumann to rule out hidden variables, didn't reject it as invalid. Von Neumann was one of the greatest mathematical geniuses of his time, so Bohr and others presumed he was right. Bell discovered, however, that von Neumann's theorem was far from complete and offered a wide range of counterexamples. The validity of Bohm's hidden variable theory, therefore, could be gauged only by different means. Bell set up to develop such a litmus test.

Before describing the theorem Bell developed, let's take a closer look at Bohm's spin state formulation of the EPR thought experiment. Recall that the idea is to hurl the two members of an entangled pair of spin ½ particles, such as electrons, in opposite directions, without knowing initially the spin states of either. Therefore, before measurement, the pair's total spin state is an equal superposition (combination) of two possibilities: "up-down" and "down-up." That is, they are anti-correlated—but we don't know yet which way. The other two options—"up-up" and "down-down"—are rejected by Pauli's exclusion principle that prohibits both electrons from being in exactly the same state.

Now let's imagine taking a measurement of the first member of the pair using a magnetic apparatus similar to that employed in the Stern–Gerlach experiment described earlier. The magnetic field can be oriented along a wide range of directions: the *x* axis, the *y* axis, the *z* axis, or some combination (for example, the *x*-*y* diagonal). Because the three components of spin—s_x, s_y, and s_z, along the *x*, *y*, and *z* directions, respectively—are noncommuting operators (meaning the order in which they are applied matters: $s_x s_y$ is not the same as $s_y s_x$), Heisenberg's uncertainty principle applies. That means if the spin value along a given direction becomes known completely (either up or down), the two other perpendicular directions have spin values that are completely indeterminate.

Remarkably, once an experimenter decides upon a magnetic field direction and observes the first electron (to be up, for instance, relative to that axis), the second one "knows" immediately to take on the opposite value along the same axis. If a second researcher lines up the second magnetic field in the same direction as the first does, she would get a definitive value that is anticorrelated. If she, in contrast, aims it in a perpendicular direction—such as if the first magnetic field lies along the *x* axis and the second lies along the *y* axis, she would get one of a random combination of two spin values—either up or down along that direction, with equal odds. It is as if the first spin value doesn't matter. Therefore, as Bohm wondered, how does the second electron have any idea how the first experimenter pointed his magnet and what result he received? Somehow, Bohm thought, along the lines of Einstein's own views on local realism, additional information needed to be conveyed between the two electrons—hence hidden variables.

In thinking over Bohm's configuration, Bell soon ruled out the "local" part of how quantum entanglement might work. Given that the particles might be separated indefinitely far away from each other before their states are measured along directions determined subsequently by the experimenters, and immediately achieve the same result, the effect cannot be local—unless somehow information is exchanged faster-than-light. As he noted in his 1964 paper, "It is the requirement of locality, or more precisely that the result of a measurement on one system be unaffected by operations on a distant system with which it has interacted in the past, that creates the essential difficulty."

However, Bell was still very open to the possibility of hidden variables that allowed the second electron's range of conceivable results for any direction of the magnet to be delineated in advance—hence, realism and determinism. To distinguish that option from orthodox quantum mechanics, he devised a kind of accounting system that would tally and compare how the results of rotating the apparatus would vary in either case. With hidden variables, all the information would be present already, meaning that a spin measurement in the *x* direction would encompass two possibilities for spin in the *y* direction: up and down. In other words, it would be additive. Those sums lead to Bell's inequality—comparisons between the tallied results of measurements along various directions that must be true if hidden variables are present.

Bell's theorem suggests a means for conducting careful experiments designed to distinguish realism—having extra information at hand, even if it isn't used—from the instant anticorrelation of quantum entanglement, which draws upon nonlocal access to a shared quantum state, yielding specific values only upon measurement. The latter seems almost magical, yet Bell's theorem (against the preferences of its designer) has consistently supported the option. Much to Bell's disappointment, no hidden variable theory has met its test.

How Light Rays Play on Opposite Days

Devising experiments with electrons regarding their spin values has practical limitations. One needs to identify two entangled electrons—for example in the ground state of a helium atom—and keep them isolated from all sources of electromagnetism and other types of interference, as they are separated and sent to two different measuring apparatuses. Such complete isolation is hypothetically possible, but not very practical, especially since there are superior ways of testing entanglement.

Far easier is using entangled photons, particles of light. Photon spin values are represented by their polarization states, either counterclockwise or clockwise, like the two different handedness types of screws. They correspond to "spin up" and "spin down" along a particular axis. Unlike neutrinos, photons can either be right-handed or left-handed. Unpolarized light is an equal mixture of both types. Polarization separates counterclockwise from clockwise states, or equivalently, horizontally polarized from vertically polarized, along a certain direction. It is very easy to polarize light—just wear a pair of polarizing sunglasses, which reduce light intensity by 50 percent, and see for yourself.

In the EPR experiment conducted with photons, unpolarized light is sent in different directions to two different analyzers that measure its polarization states. If they lie along the same line, the analyzers must record opposite polarizations—for example, if one is set to record counterclockwise, meaning "spin up," the other knows instantly to be clockwise, meaning "spin down." However, in general, the analyzers are set at angles to each other, which might vary spatially by rotating them. Therefore, Bell's inequality must be expressed in a continuous form that varies with angle and allows for mixed states, rather than the discrete *x*, *y*, and *z* components mentioned earlier. Indeed, Bell developed a continuous expression, distinguishing hidden variables from entanglement, which might readily be used in photon experiments. In particular, the correlation of the spin directions has a distinct angular dependence if standard quantum theory is correct, compared to what it would be if the hidden variables hypothesis were correct.

Practice makes perfect, even in science, so the early photon entanglement experiments were far from ironclad. One of the first loopholes identified in Bell's theorem, therefore, was that measurements always have a degree of experimental error. No apparatus is perfect. Consequently, in 1969 the physicists John F. Clauser, Michael A. Horne, Abner Shimony, and Richard A. Holt developed a generalization of Bell's theorem, known as the CHSH inequality, targeted at realistic experimental situations. It has subsequently often been used in place of Bell's inequality. Along with Stuart Freedman, Clauser conducted the first test of hidden variables versus nonlocal entanglement in 1974, based on yet another variation of Bell's inequality. While it favored the quantum explanation, the team didn't collect sufficient data for a definitive conclusion.

Another loophole physicists considered is the possibility of information reaching the detectors about their angular orientation before actual readings are made. That might happen, for instance, if the detectors' positions are fixed in place. Such advanced knowledge would spoil the plot, so to speak, by allowing the photons to know how they are going to be measured and plan their spin correlation accordingly. It is like two presenters checking with each other, several weeks before, to see what outfits they'd be wearing at the Academy Awards, but then claiming on camera during the program that it was sheer coincidence. To rectify such a possibility, the experimental settings needed to be varied quickly. Even before Bell's inequality paper, a 1957 article by Bohm and Aharonov emphasized the need for rapidly changing setups.

French physicist Alain Aspect (1947–present), known for his quantum entanglement experiments; Credit: AIP Emilio Segrè Visual Archives, with permission of Randy Hulet.

Finally, Bell and others noted a "freedom of choice" loophole, in which the experimenter's planning could subvert the goal of distinguishing hidden variables from bonafide quantum entanglement. If the researcher wasn't careful, his selections might sway the results. To cite an extreme example, if two detectors were switched periodically between certain angular positions, that pattern might lead to correlations in the data, falsely suggesting a hidden regularity in nature. To address that, experimenters have used random number generators to change the settings in an unpredictable manner.

Taking such cautionary advice into account, in 1982 the physicist Alain Aspect designed and conducted the first major Bell tests at the University of Paris in Orsay. Using lasers, polarizers, and fiber-optic cables, along with a flexible apparatus designed to be quickly changeable, their experiment won considerable praise for affirming statistically that nonlocal quantum entanglement is the correct hypothesis, by ruling out hidden variables. While Aspect's experiments were very well designed, other researchers soon pointed out that they weren't loophole-free. In

Austrian quantum physicist Anton Zeilinger (1945–present), who has conducted many pioneering experiments on quantum information, quantum teleportation, and related topics; Credit: Photographer Gerd Krizek, courtesy of IQOQI Vienna.

particular, not all of the photons were measured, leading to a "detection loophole." Since then, Aspect has continued to develop novel Bell tests with increasingly "loophole free" apparatus, making his mark as one of the key confirmers that Einstein's "spooky action at a distance" is a genuine effect, not the result of hidden variables.

Cosmic Bell Tests

Another pioneer of experimental tests of quantum foundations is the Austrian physicist Anton Zeilinger. He has made his mark in numerous important quantum teleportation experiments as well as Bell tests. Quantum teleportation is reproducing a quantum state remotely—a far cry from transporting actual materials, let alone living things, but intriguing nonetheless. Through his and his contemporaries' research, Vienna, where many important scientific findings were made in the nineteenth and early twentieth centuries, has risen again as a major physics destination.

Born in 1945, Zeilinger's youthful interests in science were spurred by his father, a biochemist. He excelled in math and physics in school, motivated by an inspiring science teacher. Attending the University of Vienna, at a time when it had a very flexible physics curriculum, he put off learning about quantum mechanics until he needed to pass a comprehensive exam. That motivated him to pick up the subject on his own—and he was hooked. As he recalled:

"Reading these textbooks, I was immediately struck by the immense mathematical beauty of quantum theory. But I got the feeling that the really fundamental questions were not addressed, a fact which just increased my curiosity."[4]

One of the many projects Zeilinger has become involved in, in recent years, involves exploiting astral light sources, such as stars and quasars, to provide loophole-free tests of quantum nonlocality. Known as cosmic Bell tests, the impetus for the experiments came from discussions with the physicists Andrew S. Friedman, Alan Guth, and David Kaiser of MIT, and others, about ways to incorporate ancient light into a modern experiment.

A group headed by Zeilinger and his collaborators, including the University of Vienna PhD students Johannes Handsteiner, Dominik Rauch, and others, conducted several groundbreaking experiments. In one project, conducted in 2017, two telescopes located more than a mile apart in Austria were aimed at two different stars. Colors drawn out from the light in each star—red and blue—were used to program the conditions for the polarizers in a Bell experiment. Because the light was hundreds of years old and generated by stars located very far away from each other, there was absolutely no chance human freedom of choice or experimental conditions would influence the experiment. The experimental results confirmed

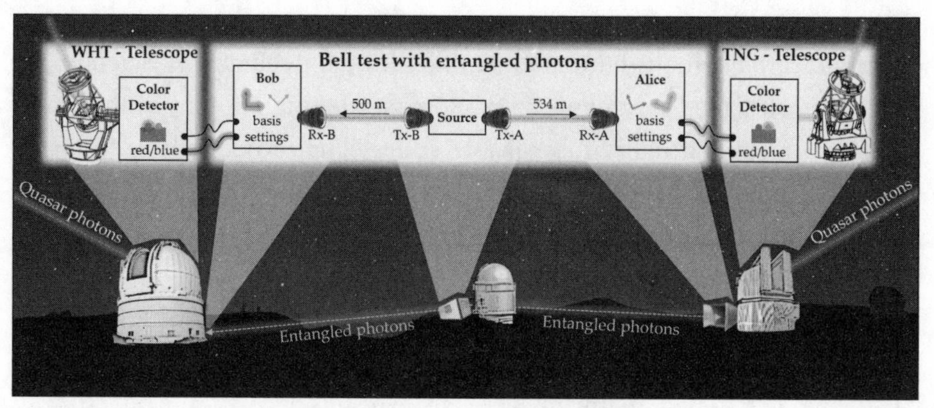

Bell Test with Entangled Photons; Credit: Dominik Rauch, IQOQI Vienna.

China's quantum science satellite Micius, passing over an optical ground station in Graz, Austria, 2017; Credit: Johannes Handsteiner, IQOQI Vienna.

that Bell's inequality was violated significantly, ruling out hidden variables more stringently than ever before. That was followed by an even more far-reaching experiment involving light from quasars billions of light-years away. Once again, Bell's inequality was breached significantly, showing how even the early universe seems to favor Bohr in his battle with Einstein.

The Zeilinger research group, with many graduate students and postdocs, has been active in numerous other projects. As part of his PhD thesis, Handsteiner was involved in a related project connected to a satellite called Micius launched by the Chinese Academy of Sciences. Micius is designed for quantum cryptography, using photons sent from space to create virtually unbreakable codes. The project points to the growing practical uses for quantum entanglement—including, for example, modeling molecules such as caffeine.

Harnessing the Quantum

Caffeine has fueled many imaginative expeditions into the quantum world, no doubt. Who knows how many innovative ideas were midwifed

Superconducting qubit chip with several cubits, used as a processor for a quantum computer; Credit: Courtesy of Michael Fang, Martinis Lab, University of California, Santa Barbara.

by a rich, dark brew at a Viennese coffeehouse? Some physicists think it's time to return the favor and model the caffeine molecule using quantum computers.

Indeed, along with presenting an intellectual mystery, quantum entanglement has an enormous potential for practical use in powerful computers. Modeling caffeine, a molecule with twenty-four atoms connected in a scaffold of chemical bonds, might sound simple—especially after a few fortifying cups of the drink for which it is named. However, because it is a quantum system, classical computers are at their limit when trying to reproduce its features.

As Richard Feynman showed in a 1981 talk, "Simulating Physics with Computers," describing quantum systems—with their various nuances such as superposition of states and entanglement—would ideally be achieved using a quantum computer. Instead of bits, 0s, and 1s of ordinary computers, their information would be stored and processed using a quantum equivalent—which has become known as qubits. These are two-state quantum units (typically particles such as electrons or photons) that exist in a superposition of both possibilities—up and down—until measured. The idea is that a quantum computer can parallel-process multiple options at once, maintaining a state of quantum coherence,

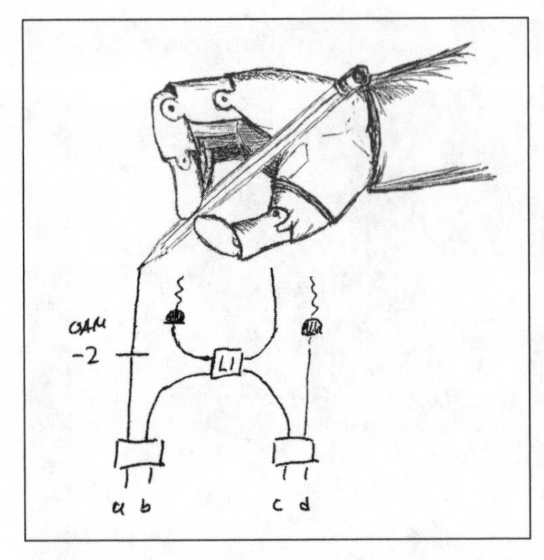

Imaginative sketch of a quantum computer replicating itself; Credit: Robert Fickler, IQOQI Vienna.

until the quantum collapse triggered by the operator demands an answer to a particular question.

Qubits have been manufactured with the goal of being used for coding and decoding, forecasting, and modeling in commercial quantum computers. Researchers estimate, for example, that to model a caffeine molecule would take a computer of 160 qubits.[5] That might not sound like much, but qubits are currently expensive and must be kept at extremely cold temperatures to be operable. Yet, given the astonishing progress with ordinary computer chips, who knows what the future will bring in terms of quantum computational progress.

It could be that future progress will be hastened by using quantum computers to design their own successors. After all, simulating quantum systems, with their layered properties, such as superposition and coherence, might be simplest by using other quantum devices. Instead of being fueled by hot beverages as they do the programming, they'd need ice-cold liquid helium (at least for qubits similar to those of today)—which unfortunately is far more expensive.

One of the most persistent notions in atomic theory, dating back to Bohr's model of 1913, is that electrons jump from one energy level to another in a sudden, instantaneous fashion, without setting any footprints in the space in between. It is a bit like passengers on a red-eye flight from

New York to London. Yes, they crossed the Atlantic, but they certainly didn't make an impression on the occupants of any ships below.

Heisenberg's matrix mechanics codified Bohr's notion of instant leaps. Schrödinger brought out wave mechanics as a more tangible alternative, with electrons climbing and sliding deliberately, not impulsively—more like well-behaved children than impetuous rascals. However, Born then countered with a probabilistic version of Schrödinger's wave functions, restoring the possibility of instant jumps, which elicited protests by its creator to no avail. By the 1950s, Schrödinger was resigned to being in the minority view in interpreting his own equation. One of the problems was that technology was not available at that time to test whether or not quantum jumps were indeed sudden and instantaneous. It was only in the 1980s that quantum jumps were first observed directly, but even then they couldn't be tracked precisely.

Flash forward to 2019, when a research team headed by Zlatko Minev, working with other scientists in the Yale physicist Michel Devoret's lab, vindicated Schrödinger's hunch by tracking an electron's leap between different energy levels and reversing it mid-flight. It matched an idea known as "quantum trajectory theory" that allows for transitions to be mapped out, like a GPS system tracing a highway journey. In the team's letter to *Nature*, they assert that their findings "demonstrate that the evolution of each completed jump is continuous, coherent and deterministic."[6]

An open question, however, is what happens if quantum jumps aren't being tracked. In quantum mechanics, observation often makes all the difference. Therefore, if an event is being watched the result might be distinct from a spontaneous event. Consequently, the jury is still out on the issue of whether or not there is a fastest rate of quantum transitions.

Ironically, while Einstein decried the notion of instantaneous interactions, his own general theory of relativity contained the seeds of such a possibility. Unlike special relativity, which places strict limits on the speeds of transportation and communication, general relativity, embracing a dynamic fabric of space-time, is far more flexible. As the Austrian mathematician Kurt Gödel demonstrated, for example, a rotating universe would permit closed timelike curves (CTCs), meaning loops that go backward in time.

Einstein's 1935 paper with Nathan Rosen, "The Particle Problem in the General Theory of Relativity," sometimes known as the "ER" paper, in contrast to the EPR paper published in the same year that included Podolsky as well, introduced the idea of connections between otherwise disparate parts of space. While Einstein–Rosen bridges, later dubbed "wormholes" by John Wheeler, are not navigable, in the late 1980s, Kip Thorne and his student Michael Morris developed traversable versions. Thorne and his group at Caltech also demonstrated how traversable wormholes, if constructed in a certain manner, could conceivably serve as CTCs and allow for passage backward in time. Under certain circumstances that might allow for retrocausality, or backward causality.

Wheeler himself was a longtime advocate of retrocausality, arguing that Maxwell's equations of electromagnetism, and other laws of nature, are perfectly time-reversible. He believed that Bohr's complementarity could take place directed into the past as well as the future—which he proposed to test using what he called a "delayed choice" experiment. An observer today might even be able to set off a quantum collapse in the past by measuring it, he believed, thus shaping the early history of the universe.

In 2012, the Australian-born Cambridge philosopher Huw Price suggested that retrocausality could logically be a feature of quantum physics—part of its built-in time-symmetry. While we don't notice reversals of the law of cause and effect in the classical world—even though its microscopic equations are time-symmetric—he argued that retrocausality might be necessary to explain certain perplexing aspects of quantum mechanics, such as its apparent nonlocality.

Extending Price's hypothesis, in 2017 the researchers Matthew Leifer of Chapman University and Matthew Pusey of the Perimeter Institute constructed a way to employ retrocausality to restore realism to descriptions of quantum entanglement. Hence, instead of assuming instant nonlocal correlations between quantum states, information could be transferred backward as well as forward in time in a manner consistent with observation. The researchers argue that in the block universe of general relativity, in which space-time is a unified construct, quantum mechanics needs to encompass the past, present, and future in a seamless fashion. Such would be possible by reconciling quantum theory with the timeless reality sketched by Einstein in his description of the universe.

A fundamental theory of nature would ideally combine quantum field theory and general relativity in a seamless fashion. Merging those disparate constructs, however, would be in some ways akin to the adjustments made by a couple moving in together. Not every aspect of each could be maintained in the same way.

Two single people might start out with completely different lifestyles and expectations. One might imagine living in an apartment high above a city, with an amazing view. The other might picture a ranch house with rolling land surrounding it as far as the eye can see. Vacations for one might be cruises on the open sea, while for the other it might be volunteering on an alpaca farm in the Andes. Yet if they decided to meld their lives, they might well need to compromise.

Similarly, for quantum physics and general relativity, a completely unified description of nature would require a novel set of ground rules. For example, should the basis of the union be in ordinary space or abstract Hilbert space? Does reality need to be based on the physical dimensions we observe and measure? Or might it be the case that Hilbert space, as the space of quantum states, is somehow more fundamental?

Let's start with the supposition that physical space-time, as described by general relativity, is real. How, then, to explain long-range correlations such as entanglement? Could the malleability of space-time, including the possibility of shortcuts such as wormholes, lead to a network of hidden connections?

In 2013, the noted physicists Leonard Susskind of Stanford and Juan Maldacena speculated if ER and EPR could somehow be linked; that is, if wormholes could somehow act as channels for entanglement. While Einstein never drew such a conclusion, the notion that his two major contributions from 1935 are deeply connected is a fascinating one.

Susskind and Maldacena imagine how quantum teleportation and entanglement might work by using wave functions threaded through wormholes to explain such interconnections. They speculate that the universe could be pockmarked with myriad Planck-scale wormholes, too minuscule to observe, which through their interconnectivity could explain all manner of quantum phenomena. It is certainly an intriguing thought.

Conversely, other theorists have attempted to start from pure Hilbert space, and generate physical space-time as a consequence. In such notions, quantum entanglement is fundamental, but actual physical connections, including the laws of general relativity, are derivative. Einstein, if he were alive today, would be shocked, no doubt, to learn that quantum physicists are trying to subsume general relativity into their theory, just as he tried unsuccessfully to do with theirs.

As the quantum physicist Časlav Brukner noted:

> I think that coarse-grained observables, and symmetries might indicate that some "geometrical" notions might be derivable from abstract quantum theory. Is it possible to arrive at notions of nearness, distance, and space—and, furthermore, at the theories referring to these notions, such as the theory of relativity, quantum field theory, and elementary-particle theory—merely on the basis of clicks in detectors? Or is it necessary to presuppose these notions, prior to the construction of physical theories? To me, this is one of the most pressing contemporary questions in the foundations of quantum mechanics.[7]

Truly, humanity has developed a much richer understanding of how the universe is interconnected since the days of Empedocles and Pythagoras, and even since the time of Einstein and Bohr. Yet there is still much work to be done in explaining why certain connections transpire at maximum speeds, while others appear to coordinate immediately. Though Pauli and Jung's dialogue was not purely scientific, they did identify a significant dichotomy in nature: the distinction between causal linkages and synchronous connections. If a unified principle could explain both, conceptual progress would surely be made.

CONCLUSION

Unraveling the Universe's Tangled Web

HERE WE STAND, ON A DIMINUTIVE PLANET, ORBITING A MIDDLING star, a glowing candle peripherally perched in one of myriad whirling galactic chandeliers scattered among an unimaginably vast expanse of darkness. We are manifestly alone. Isolated in space, we are also dwarfed by the immensity of time. Our limited stay on Earth pales in comparison to the roughly 13.8 billion years since the Big Bang. Yet despite such limitations, we brazenly hope to fathom as much of the colossal scope of reality as humanly possible.

Since ancient times we've gazed at the nighttime sky searching for wisdom and connection. In the age of Homeric heroes, we imagined a vibrant interplay between the celestial figures and us mortals. Thousands of years later, Galileo's telescope offered a glimpse of the colossal distances involved. He speculated that while light traveled swiftly, it had a finite speed, meaning that sunlight takes some time to reach us and starlight far more so. When Michelson's measurements pinned that speed down, and Einstein demonstrated that it was the upper limit of communication in conventional space, our remoteness became much more apparent. Einstein's general theory of relativity heralded the dawn of the age of mathematical cosmology, which has revealed a vast, expanding cosmos—with an observable extent tens of billions of light-years across. How naive it was back in earlier times to purport to interact with beings in the heavens—to imagine, for example, that the constellation Andromeda, in which an entire galaxy (M31 or the Andromeda galaxy) is centered (according to our line of sight, but much farther away than the stars in that constellation), was once an Ethiopian princess chained to a coastal rock, as Greek mythology held.

Classical science made great strides by embracing causality—the notion that reality is like a chain of toppling dominoes, each pressing on its neighbor and compelling it to fall. Yet, as quantum physics shows us, mechanistic causality, as natural as it seems, is not enough to explain the wide range of phenomena in the universe.

The Limits of Reason

Humans are expert puzzle solvers. Our brains look for patterns in the world around us. We soak in information through our senses and instruments. We process that assembled data and naturally make certain associations. Ideally, the links we forge in our minds help us anticipate the perils and opportunities lying ahead, enabling us to make superior choices.

Our pattern recognition is not always an accurate representation of reality, however. A significant portion of sensory associations are meaningless—the chance connection of two pieces of information that happen to arrive around the same time. For example, suppose we are walking down the corridor of an old office building and happen to notice a rattling sound from clunky air ducts. We might start to imagine that the clanging of the ducts is timed with the beat of our footsteps in some kind of strange symphony. Or, while making our way down the same hallway, we might happen to observe a musty scent hitting our nostrils as we avoid tripping over a bumpy, raggedly carpeted part of the floor.

Returning to that facility again and again might cement those connections. As soon as we entered, our brains might fixate on particular sights, sounds, and odors. Nevertheless, those reactions might be spurious correlations—things that seem to be connected but really aren't. For example, the musty smell could be due to fumes from a nearby factory seeping in through the window, rather than from the flooring. Or it could be that we are so nervous and drenched with perspiration by the time we reach that dreaded bump, and primed to pick up unusual scents, we can sense the result of our own fear. In a ghastly feedback loop, we might focus inordinately on what we dread.

Why are our brains adept at forming associations? Survival sometimes depends on our ability to make connections. The stench of rotting food rightly triggers us not to want to eat it. Loud car horns spur us to run out of the way. These reactions might save our lives. No wonder our brains naturally forge such links.

While disagreeable sensations might lead to avoidance, pleasant ones might invite many happy returns. Many department stores and shopping malls make use of that principle by perfuming their air and piping in upbeat music, hoping to produce smiles and sales. When we pack the stuffed shopping bags into the trunks of our cars, we might not realize that it was the lilting voices and lavender that primed us to purchase.

Science functions by looking for actual, meaningful, reproducible connections—not just coincidences. To correctly anticipate the future, we develop models that distinguish genuine relationships from spurious correlations. Mere "gut feelings" that two things are linked can be misleading. That's why potential medications are often tested against placebos to make sure of their efficacy. Just because someone thinks a product works the way it is supposed to doesn't mean it actually does. Similarly, that's why crime statistics are a much better measure than reading disturbing stories in the news and reaching unwarranted conclusions about safety threats. Our brains are trained to look for patterns and sometimes create them when they aren't really there. In probing the mysteries of the universe, we need to keep in mind the danger of such mirages.

Psychological studies have shown that subjects faced with random data often see patterns if they anticipate that they might be there. For example, an experiment conducted by John Wright in 1962 asked subjects to press a circular panel of sixteen push buttons in the "correct" sequence to obtain a reward. While in actuality, the rewards were doled out with no rhyme or reason no matter how the buttons were pressed; whenever subjects were rewarded frequently (say, 80 percent of the time) they tended to begin to prefer certain orderings.[1] In other words, they developed a "superstition" that particular sequences of buttons allowed them to win. Another psychological experiment, conducted by Harold Hake and Ray Hyman in 1953, similarly demonstrated how subjects tend to seek order in chance progressions. It detailed how past successes in guessing random sequences of horizontal and vertical rows of light bulbs skewed subjects' predictions about which direction of bulbs would appear next.[2] One lesson to be learned from these experiments is that the same pattern recognition skills that allow us to do true science can also lead to unwarranted conclusions and even superstitious behavior.

More recently, the Princeton astrophysicist Lyman Page was giving a talk about cosmic background radiation data collected by the WMAP (Wilkinson Microwave Anisotropy Probe) space probe. While projecting an image of a sky map of that information, he happened to notice the initials "S H" embedded in an otherwise nondescript region. Page jokingly remarked that the formation was the Cambridge physicist Stephen Hawking's engram in space.[3] He then clarified that in any set of data, the human mind is bound to discover meaningless patterns. Good statistics (for example, models based on the work of the English statistician Thomas Bayes) should aspire to weed out such prejudices in favor of actual connections.

Two of the most startling natural sensations are the flash of lightning and the boom of thunder. Anyone who has experienced thunderstorms knows that the boom typically follows the flash. Light waves, we know, travel far more quickly than sound waves: about 186,000 miles per second versus roughly one-fifth of a mile per second. Compared to the latter, the former might seem instantaneous. However, science has well established that light, though incredibly swift, travels at a finite speed.

Pattern solving shows its efficacy when dealing with storms. By clocking the delay between the sight and the sound, science informs us that we can gauge how close an electrical storm lies. Therefore, an especially short interval between the two signals would indicate that the audible boom didn't have to journey far, pointing to a dangerously close storm cloud. For instance, less than a thirty-second delay would point to electrical activity within six miles. We'd then be wise to look for shelter. In short, a vivid scientific model of the connection between lightning and thunder produces a greater potential for safety.

But thunderstorms also produce emotional reactions that might lead to false conclusions. Even if they aren't a true threat, they can be very scary and lead to unwarranted anxiety. The sudden onset of turbulent weather triggers such primal terror that it has become a staple of horror flicks. No wonder the ancients considered them bad omens. In some cultures, they epitomized the wrath of the gods. Before the speed of light and the nature of electrical storms were established, hapless folks might worry that a thunderbolt could arrive simply out of the blue—an instant burst of anger from a god such as Thor. Even in modern times, they seem to trigger instinctive feelings that something horrible might happen.

Associating lightning, thunder, and electricity is science. Believing they offer premonitions of evil (aside from the possibility of downed branches, electrocution, and other direct physical consequences) is pseudoscience. Modern statistical methods, applied to data about stormy weather and ill fortune, would distinguish the valid relationship from the false hunch.

Reasoning doesn't always guide feelings, though. If right after one memorable thunderstorm, a woman received news that her husband died in a freak accident, and then right after another storm two years later she got word that her son perished in a horrible motorcycle crash, she might start to believe that the meteorological events were harbingers of bad luck. It might be hard for a statistician to convince her otherwise. Such is the persistence of our personal chance associations, especially those that are extremely emotive.

Clearly, our capacity to make connections—with its dogged automaticity and deep association with strong emotions—can lead us to scientifically valid conclusions or can lead us astray. It can offer credible predictions—such as in the case of measuring the proximity of a storm—that enable us to avoid actual dangers or otherwise improve our lives. Or it can lead to superstition and phobias—some innocuous, others incapacitating. The boundaries between which kinds of connections are considered real (demonstrable through science) and which are considered supernatural (spiritual, faith-based, or pseudoscientific, depending on one's perspective) have shifted significantly throughout recorded history. For example, from ancient times until relatively recently in history, the professions of astrology and astronomy were closely intertwined. Sagacious stargazers who offered predictions about the motions of the planets and constellations also offered horoscopes. Powerful figures would rely on such astrological forecasts to plan their worldly lives.

The late astronomer Carl Sagan was wont to point out how silly it was to believe that astral arrangements affected human traits. After all, he noted, the medical professionals gathered around a mother when she gives birth have demonstrably far more of a gravitational influence on the baby—albeit still minuscule—than do distant astral bodies such as the stars of Sagittarius. If it were nighttime and the hospital had a roof, the dim light of such bodies wouldn't even be able to penetrate the building and affect the newborn. If it were daytime (and not a solar eclipse), any starlight or reflected light by other planets would be vastly outshined by the Sun. Consequently, there would be no logical way the particular alignment of stars and planets at the time of birth could guide the child's fate—or how astral influences could affect the fate of anyone for that matter.

For rational thinkers, astronomy and astrology started to diverge around the time that the seventeenth-century English physicist Isaac Newton advanced a remarkably accurate description of physical reality that clearly distinguished the former from the latter. He modeled the orbits of planets, along with an extraordinarily wide range of other natural phenomena, by means of a powerful set of mechanical laws. In tandem, Newton's laws of motion and Newton's law of universal gravitation showed how genuine influences between objects—physical forces—could act over a broad span of distances to cause predictable behavior. Newtonian mechanics could pinpoint astronomical events, such as the periodic alignment of Mercury and Venus in their orbits around the Sun, with absolute precision. A network of causal connections—defined with mathematical precision—linked all objects in the cosmos, it suggested. In contrast, no reproducible model could show that babies born when such planets appear close together in the sky would grow up with certain characteristics. Hence, among those rational thinkers who valued scientifically proven connections and rejected speculative influences, interest in astronomy blossomed and passion for astrology withered.

Why weren't the laws of motion discovered earlier? On face value, the true workings of forces are not at all obvious. The rudimentary understanding most of us acquire early in childhood does not naturally lead to the richer picture that Newton proposed.

By the age of eight months most babies have developed a sense that pushing or pulling on an object makes it move. They start to notice, at various early stages of development, that they and others can affect motion through exertion—intentional or otherwise. Then, in a naughty moment, many babies nudge their food off their high chairs, watch it drop to the floor in a goopy mess, and observe their parents' looks of horror, pride, or a mixture of both, as the case may be. They've learned that a push or pull can produce a reaction.

Brukner has remarked how that stage of child development corresponds to the realization of causality, the notion that causes produce effects:

"Causal thinking spontaneously arises in a child at about the time when she or he realizes that by exerting forces on nearby objects, the child can make these objects move according to their will. Causal relations are revealed by observing what would happen in the world (for instance, with the child's object) if a given parameter (the child's will) were separated out from the rest of the world and could be chosen freely."[4]

If all scientific phenomena were causal, we could dismiss anything that lacks a mechanism for cause and effect. Yet quantum phenomena such as entanglement clearly are a different breed. They are scientific, in the sense of reproducibility and accurate predictions, but are manifestly acausal. Our instincts from childhood would lead us astray. We have to think beyond our formative beliefs, in order to fathom (as much as possible) the quantum world.

Nature is indeed a crafty gamer. Its mechanisms range from simple, direct rules to subtle, hidden connections. To match its shrewd moves, humanity has been forced to think like a master player. In a tournament of inquiry lasting thousands of years, Nature has been adept at keeping countless philosophers, theologians, scientists, and other thinkers on their toes. Frustratingly, it appears to have distinct rule books for different kinds of processes and different scales. Whether or not science will someday identify a universal playbook that explains everything from the extremely small to the extraordinarily large remains an open question. In the meanwhile, as various thinkers have grappled with interpreting new findings, flexibility has been the key to advancement. Those who play a variety of games typically maintain an arsenal of different strategies for success. Seasoned gamers learn how to adjust to rules that vary from one pursuit to another. Scientists have discovered that they need to do the same.

One of humanity's aspirations is to explore the universe. In trying to envision space missions beyond our solar system, we encounter overwhelming obstacles due to the enormous distances involved. If all communication and transportation were instantaneous, such exploration wouldn't present much of a problem. But the light-speed limitation renders that prospect daunting. To progress, we need to think flexibly and try to develop solutions to the distance hurdle.

Beyond the Limits of Light-speed Causality

Ironically, though Einstein's special theory of relativity suggested that the vacuum speed of light set an absolute upper bound on interactions, his subsequent general theory offered a significant loophole: space-time is malleable. Moreover, the development of quantum mechanics soon thereafter opened up another crack in that ceiling: acausal correlations (or anti-correlations) might transpire faster than causality's limits. In tandem, the warpability of space-time, and long-range quantum connections such as entanglement and coherence, offer the prospect that future interstellar civilizations might find ways to circumvent the light-speed limit.

Even more far-reaching scenarios, such as the existence of tachyons or higher dimensions, might potentially allow additional avenues for supraluminal interactions. While those notions sound incredible, laboratory scientists at CERN and elsewhere take them very seriously. Mathematically, many theorists find them too enticing to rule out. Nevertheless, consistent null results in numerous experiments to test such hypotheses would eventually quash all hope.

An important subset of higher-dimensional speculation lies in the hypothesis of "brane-worlds." Brane-world scenarios were originally developed in the late 1990s by Nima Arkani-Hamed, Savas Dimopoulos, Gia Dvali, Ignatios Antoniadis, Lisa Randall, Raman Sundrum, and others as variations of the class of unification models known as string theory and M-theory.[5] Such models replace point particles with vibrating strands, loops, and sheets of energy, residing and interacting in a higher-dimensional manifold (generalization of space-time). Three of those dimensions are familiar space, one is time, and the others are supposed to be curled up in tight balls or knots so that they are effectively unobservable. However, in brane-world scenarios, one of the extra dimensions is large enough that it might potentially be probed. That leads to the notion of making use of that extra dimension in field theory to resolve certain persistent conundrums, such as the fact that gravitation is much weaker than all of the other forces.

To explain the weakness of gravitation, in the Randall–Sundrum brane-world scenarios, the universe we observe is confined to a three-dimensional spatial membrane, but gravitons (the exchange bosons conveying gravitation—analogous to photons conveying electromagnetism) are able to leak into an extra-dimensional region beyond it, called the "bulk." No other forces or particles might traverse the bulk. Rather, they are stuck to the "brane" (our three-dimensional membrane). The leakage of gravitons explains why gravitation is so much weaker. It is just like if multiple faucets emptied into a sink, but one was attached to a pipe with a leak, the water pressure of the stream from the leaky pipe might be much lower. Similarly, the passage of gravitons into the bulk dilutes the force of gravity compared to the strong, weak, and electromagnetic interactions, which are confined to the brane.

Since 2015, we now know the gravitational waves are real. Starting then, the LIGO (Laser Interferometer Gravitational-Wave Observatory) detectors have been picking up signals from remote catastrophic events in space, such as black hole collisions. Conceivably, if brane-world scenarios are real, advanced civilizations in the distant future might find a way of modulating gravitational wave pulses and finding a means for them to take shortcuts through the bulk to remote parts of space. As in the case of wormholes, that might conceivably allow for supraluminal communication.

Finally, and even more speculatively, it might be the case that the space-time itself is not fundamental. In some scenarios, abstract Hilbert space would be basic and geometry emergent. In that case, the light-speed limit would be a secondary property that developed as the outcome of quantum rules, emerging perhaps through optimizing among a set of conceivable transitions. In other words, among the set of all possible transactions in Hilbert space that correspond to what we would call particle interactions in ordinary space, perhaps those transpiring at light speed would have some advantage. We already know what that advantage is in conventional space-time—light journeys through space in a way that optimizes the time traveled; that is, it takes the path of least time. We would need to find an analogous reason in Hilbert space that would produce the same effect. If so, there might be parts of Hilbert space in which faster communication through space-time turns out to be optimal. In that case, maybe we could exploit those alternatives to allow for supraluminal communication and travel.

All this sounds extremely abstract. However, in Vienna, the city of Pauli's birth, considerable progress is being made by researchers at the IQOQI, including Zeilinger, Brukner, and others, in trying to understand the nature of causality and information transfer in quantum systems. Such research is a fitting tribute to the legacy of Pauli, who worked hard to show how quantum mechanics, with its emphasis on acausal relationships (in tandem with deterministic equations) and on the essential role of observers, differed from classical theories that were purely mechanistic.

The Legacy of the Pauli–Jung Dialogue

Pauli was far from the only thinker who brought symmetry principles into physics. Noether, Hilbert, and Weyl—brilliant minds of Göttingen—deserve much of the credit. Heisenberg's introduction of the concept of isospin in 1932 did much to promote the idea. Later, the Hungarian physicist Eugene Wigner, the American physicist Murray Gell-Mann, and others picked up the gauntlet.

Pauli was similarly not the only physicist to emphasize the critical function of observers in quantum mechanics. In that domain, he drew from the worldview of Bohr. Later, John Wheeler would become the leading advocate of exploring the symbiosis between the conscious measurer and the quantum systems being measured.

Yet, uniquely, during his heyday, Pauli commanded immense respect from his peers—many of whom considered him the final arbiter of any question. Even Einstein sought out his opinion. If Einstein was the honored king, Pauli was the authoritative supreme judge. When he nodded his head in confirmation during any seminar, the speaker breathed a sigh of relief. On the other hand, if he dismissed an idea, he was probably correct in his criticism (with his initial dismissal of quantum spin a rare exception to the rule). Pauli's embracing the notion of reframing modern physics in terms of novel fundamental principles, such as CPT invariance and other applications of symmetry groups, and continuing to advocate the unification of natural forces—while moving cautiously, and taking care not to hype the efforts—resonated with the physics community in general and established a precedent for future endeavors. The decades following Pauli's death would be the most innovative period for applying symmetry principles to unification—leading to the Standard Model of particle physics.

Jung's role in all this should not be underestimated. True, his theory of archetypes and the collective unconscious, while creative and compelling, has never been scientifically proven. There is absolutely no evidence that symbolism in dreams relates to some inherited primal pattern. If one studies Eastern philosophy, alchemy, and the occult, naturally some of the icons of those subjects, such as mandalas and alchemical symbols, might be more likely to pop up in nocturnal imaginations. Even if they don't, the waking mind might reinterpret imagery to make that connection. Yet the same might be true for those whose obsessive hobby is collecting comic books; they might dream about superheroes and villains. However, in analyzing Pauli's ruminations, Jung tapped into a rich vein of brilliant speculation about principles to organize nature. His analysis served as a feedback loop that helped to amplify Pauli's ideas, while enhancing his own rudimentary speculations about how nature is connected.

Jung's synchronicity idea, as further shaped by Pauli, might thereby be seen not only within the context of psychology, where evidence is purely anecdotal and thereby lacks the validation of controlled scientific studies, but also in a call to address the revolutionary changes spawned by quantum mechanics with novel ways of understanding how the cosmos is connected. Beyond absolute causality and pure chance lies *terra incognita*: the universe's network of acausal connections. By setting the notion of synchronicity on par with causality, therefore, Pauli and Jung rightly pointed out that a unified theory of reality must account for acausal relationships, including symmetry considerations, as well as deterministic rules, in a manner that ultimately explains both.

Serendipity versus Science

Demanding scientific and philosophy concepts are sometimes sugar-coated for digestion by popular culture. In many cases, that's not as bad as it sounds—not everyone might find pure mathematical or philosophical discourse palatable. A dash of sweetener might make the factual explanations more fun. A diet of pure sweets, on the other hand, is not particularly healthy—which in this case would be misrepresenting genuine acausal correlations with common serendipity, perfectly explainable by chance encounters and selective memories.

Indeed, synchronicity has entered the cultural lexicon as an emblem of unexpectedly related events that arise simultaneously or at least in close succession. In its colloquial meaning, the term might arise when two friends blurt out similar things at the same time, or exchange near-identically themed emails that happen to cross in the electronic cloud. If two classmates arrive at a dance both wearing blue outfits, that's not synchronicity in the Jungian sense, let alone a miraculous occurrence, but rather just a simple coincidence. Still, they might call it synchronicity, as an expression of surprise by the matchup.

The term "synchronicity" was widely popularized during the 1970s through books such as Arthur Koestler's *The Roots of Coincidence*, which discusses Jungian theory within the context of parapsychology. Bringing up the emergence of quantum nonlocality, he tries to make a case for the scientific appreciation of extrasensory perception and the paranormal. However, the book was criticized for not distinguishing genuine science, with its rigorous methods, from pseudoscience. As Wolfe Mays writes in his book review, published in the *Journal of British Society of Phenomenology*:

> Koestler makes no attempt to relate ESP and quantum physics: at the most he shows that both have to use concepts which are foreign to those of common sense and Newtonian physics . . .
>
> An essential difference between the phenomena discussed in subatomic physics, for example, the behaviour of physical particles in the cloud chamber, and parapsychological phenomena is that in the former predictions can be made from mathematical theories which can then be tested experimentally by any competent physicist.[6]

In his book, Koestler proposes a "filter theory" purporting to explain why, if telepathy is possible, we normally block out the thoughts of others and focus on our own. A far-fetched conjecture indeed, with absolutely no basis in neuroscience. Much more reasonably, we might turn that idea on its head and explain seemingly meaningful coincidences because our minds tend to filter out information that does not form memorable kinds of patterns. Thus, if classmates arrive at a prom wearing different-colored outfits, that might be as memorable as if they happened to wear the same.

Many frequent travelers, for example, have experiences running into their neighbors or friends at remote airport lounges or railway stations. They're much more likely to recall such unexpected coincidences than all the times they sat in lounges with random strangers. All those non-patterns would be filtered out, offering the illusion of an astonishingly rare encounter.

In 1987, F. David Peat, a close collaborator of Bohm's, wrote a book and several articles about synchronicity, contributing to the idea's popularity, especially in connection with quantum physics. By that time,

Bohm had advanced a notion about an underlying web of hidden connections in nature, called the "implicate order." Peat showed how Bohm's idea had commonalities with synchronicity.

Arguably the most prominent public rendition of synchronicity was when the Police, a rock band that sometimes drew upon philosophical motifs, released a chart-topping album of that name in 1983. It included the songs "Synchronicity I," a musical ode to the concept, and "Synchronicity II," a catchy hit that used the Loch Ness Monster as a metaphor for submerged influences arising at key moments. The band's lead singer and songwriter Gordon Sumner, known professionally as Sting, explained his motivation for writing the latter, which still receives considerable airplay: "This song was an attempt to link a tale of suburban alienation with symbolic events at a distance, i.e., the monster emerging from a Scottish lake and a domestic melodrama. I was trying to dramatize Jung's theory of meaningful coincidence, but it was a rocking song nonetheless!"[7]

The well-read Sumner, who studied to be a schoolteacher before achieving rock-and-roll success, further explained that he came across Jung's work in Koestler's account. Rock critics gave *Synchronicity*, which turned out to be the final studio album for the Police, high praise for delving into the notion of acausal connections. As Christopher Connelly wrote for *Rolling Stone* magazine:

"Indeed, it is a crystallized expression of synchronicity itself: the idea that things that are linked need not have direct, causal relationships; that we need not resolve the distinctions that separate us to be related in an important way. 'A connecting principle/ Linked to the invisible . . . ,' Sting sings in 'Synchronicity I,' and what that connecting principle amounts to is human will and intelligence: the ability to tolerate—even celebrate—differences, while striving toward unity and understanding."[8]

Cautiously Embracing the Acausal

While avoiding the selection error of assuming every coincidence represents a veritable acausal connection, we need not veer in the opposite direction and stick only to purely mechanistic, causal explanations for all of science. Bell's inequality, and modern methods in quantum measurement theory, show how to pursue long-range acausal connections scientifically. Wormholes, and the complex connectivity of space-time in general relativity, show that even in non-quantum physics, we must address the possibility of systematic violations of the notion that cause must precede effect.

Proceeding steadily, but cautiously, physicists might aspire to craft universal rules that embrace both causal and acausal connections—perhaps in rewriting general relativity by means of quantum entanglement, or conversely by somehow exploiting the malleability of the former to explain the latter. The quest of Einstein, Heisenberg, Pauli, and others for a unified theory of nature might someday be realized—in a manner, perhaps, that twenty-first-century science wouldn't even expect.

So, if you find yourself riding a superconducting maglev train, hovering above the tracks as it flies past beautiful scenery in Japan or elsewhere, and you doze off and dream about Zen Buddhist symbols only to see one when you open your eyes and look at a design element in the carriage, you might rightly call out "synchronicity." Not about your dream, mind you. That's just a simple coincidence. Rather, the synchronicity lies in the myriad Cooper pairs of electrons, lined up in a coherent quantum state, driving the magnetic repulsion that significantly reduces the train's friction. Explain that to the person on the seat next to you, and they'd be suitably impressed.

Modern physics is strange, no doubt. However, as long as it leads to reproducible results, it offers a magnificent tower of knowledge that stands miles above the mere speculations of our ancestors.

ACKNOWLEDGMENTS

I WOULD LIKE to thank the faculty, staff, and administration of the University of the Sciences for their generous support, particularly Paul Katz, Elia Eschenazi, Vojislava Pophristic, Elisabeth Morlino, Grace Farber, Jean-Francois Jasmin, Phyllis Blumberg, Charles Myers, Leslie Bowman, Matthew Gallagher, Jon Drucker, Peter Miller, Suzanne Murphy, Tricia Purcell, Sam Talcott, Kevin Murphy, Lia Vas, Salar Alsardary, Ed Reimers, Michael Robert, Amy Kimchuk, Carl Walasek, Laura Pontiggia, Peter Kim, Abolfazi Saghafi, Tarlok Aurora, Bernard Brunner, Sergio Freire, Jessie Taylor, and Roberto Ramos. Many thanks to my outstanding editor, TJ Kelleher, for his helpful editorial suggestions and to my excellent agent, Giles Anderson, for his support. I appreciate the exceptional resources of the American Institute for Physics Center for the History of Physics, including its valuable Oral History Archive, as well as the support of the APS History of Physics Forum, the APS Historic Sites Committee, the American Philosophical Society, and the Science History Institute.

I am immensely grateful to Freeman Dyson, Stanley Deser, Kurt Gottfried, Alan Chodos, David Cassidy, Alberto Martinez, Časlav Brukner, Diana Kormos-Buchwald, Karl von Meyenn, John Donoghue, and the late John Smythies, for their fascinating insights, historical references, and recollections.

Many thanks to Bryn Mawr College Library, the University of California, Santa Barbara, the American Institute of Physics Emilio Segrè Visual Archives, Randy Hulet, Jochen Heisenberg, Kurt Gottfried, Anton Zeilinger, and the Institute for Quantum Optics and Quantum Information–Vienna (IQOQI), for generously supplying the images for this book and for the kind permission to use these images. Thanks also to Anita Hollier and the CERN Archive, and to Claudia Briellmann of the ETH-Zürich Library and Archive, for productive visits to view the correspondence of Pauli, Jung, and others.

I greatly value the support of the history of physics and science writing communities, including Gregory Good, Joseph Martin, Robert Crease, Peter Pesic, Cameron Reed, Catherine Westfall, Roger Stuewer, Gerald Holton, Stephen Brush, John Heilbron, Virginia Trimble, Paul Cadden-Zimansky, Michal Meyer, Mark Wolverton, Amanda Gefter, Dave Goldberg, Cormac O'Raifeartaigh, Konrad Kleinknecht, Thony Christie, Ash Jogalekar, Lautaro Vergara, David Schwartz, Nicholas Booth, Frank Cross, Marcus Chown, Graham Farmelo, Cortney Bougher, Kathleen Damiani, and others.

I appreciate the kind guidance of Hayden Sando, Joshua Kroll, Patrick Pham, Jonathan Caughey, Elizabeth Shanefield, Benjamin Hoffmann, and others in helping me achieve focus and creativity in diverse pursuits.

Thanks also to my friends and family, including Michael Erlich, Fred Schuepfer, Pam Quick, Simone Zelitch, Doug Buchholz, Ben Hinerfeld, Bob Jantzen, Lisa Tenzin-Dolma, Mitchell and Wendy Kaltz, Mark Singer, Nikki McGeary, Scott Veggeberg, Marcie Glicksman, Karl and Dori Middleman, Jeff Shuben, Meg and Woody Carsky-Wilson, Debra DeRuyver, Dan Tobocman, Bob and Karen Huber, Kris Olson, Shara Evans, Lane Hurewitz, Jill Bernstein, Jerry Antner, Shawn and Charlotte Williams, Richard and Anita Halpern, Jake Halpern, Emily Halpern, Alan and Beth Halpern, Tessa Halpern, Ken Halpern, Aaron Stanbro, Thessaly McFall, Arlene and Joseph Finston, Stanley Halpern, and above all my wife, Felicia, and sons, Aden and Eli, who have each offered many helpful ideas and insights over the years.

FURTHER READING

Ananthaswamy, Anil, *Through Two Doors at Once: The Elegant Experiment That Captures the Enigma of Our Quantum Reality* (New York: Dutton, 2018).

Baggott, Jim, *The Quantum Cookbook: Mathematical Recipes of the Foundations for Quantum Mechanics* (New York: Oxford University Press, 2020).

———, *The Quantum Story: A History in 40 Moments* (New York: Oxford University Press, 2011).

Ball, Philip, *Beyond Weird: Why Everything You Thought You Knew About Quantum Physics Is Different* (Chicago: University of Chicago Press, 2018).

Becker, Adam, *What Is Real? The Unfinished Quest for the Meaning of Quantum Physics* (New York: Basic Books, 2018).

Bernstein, Jeremy, *Quantum Profiles* (Princeton: Princeton University Press, 1991).

Byrne, Peter, *The Many Worlds of Hugh Everett III: Multiple Universes, Mutual Assured Destruction, and the Meltdown of a Nuclear Family* (New York: Oxford University Press, 2013).

Carroll, Sean M., *Something Deeply Hidden: The Unspeakable Implications of Quantum Reality, from Spooky Action to Many Worlds* (New York: Dutton, 2019).

Cassidy, David, *Beyond Uncertainty: Heisenberg, Quantum Physics, and the Bomb* (New York: Bellevue Literary Press, 2009).

———, *Uncertainty: The Life and Science of Werner Heisenberg* (San Francisco: W. H. Freeman & Co., 1991).

Clegg, Brian, *Light Years and Time Travel—An Exploration of Mankind's Enduring Fascination with Light* (Hoboken, NJ: Wiley, 2001).

Close, Frank, *The Infinity Puzzle: Quantum Field Theory and the Hunt for an Orderly Universe* (New York: Basic Books, 2013).

Crease, Robert P., and Alfred S. Goldhaber, *The Quantum Moment: How Planck, Bohr, Einstein, and Heisenberg Taught Us to Love Uncertainty* (New York: W. W. Norton & Co., 2015).

Crease, Robert P., and Charles C. Mann, *The Second Creation: Makers of the Revolution in Twentieth-Century Physics* (New Brunswick, NJ: Rutgers University Press, 1996).

Davies, Paul, *The Cosmic Blueprint* (New York: Simon and Schuster, 1988).

Enz, Charles P., *No Time to Be Brief—A Scientific Biography of Wolfgang Pauli* (New York: Oxford University Press, 2002).

Farmelo, Graham, *The Strangest Man: The Hidden Life of Paul Dirac, Mystic of the Atom* (New York: Basic Books, 2009).

———, *The Universe Speaks in Numbers: How Modern Math Reveals Nature's Deepest Secrets* (New York: Basic Books, 2019).

Feynman, Richard P., *QED: The Strange Theory of Light and Matter* (Princeton, NJ: Princeton University Press, 1985).

Fine, Arthur, *The Shaky Game: Einstein, Realism and the Quantum Theory* (Chicago: University of Chicago Press, 1986).

Gieser, Suzanne, *The Innermost Kernel—Depth Psychology and Quantum Physics: Wolfgang Pauli's Dialogue with C.G. Jung* (Berlin: Springer, 2005).

Greenblatt, Stephen, *The Swerve: How the World Became Modern* (New York: Norton, 2011).

Halliwell, J. J., J. Perez-Mercader, and W. H. Zurek, eds., *The Physical Origins of Time-Asymmetry* (Cambridge: Cambridge University Press, 1996).

Halpern, Paul, *Einstein's Dice and Schrödinger's Cat: How Two Great Minds Battled Quantum Randomness to Create a Unified Theory of Physics* (New York: Basic Books, 2015).

———, *The Pursuit of Destiny: A History of Prediction* (Cambridge, MA: Perseus, 2000).

———, *The Quantum Labyrinth: How Richard Feynman and John Wheeler Revolutionized Time and Reality* (New York: Basic Books, 2017).

———, *Time Journeys: A Search for Cosmic Destiny and Meaning* (New York: McGraw-Hill, 1990).

Herbert, Nick, *Faster Than Light: Superluminal Loopholes in Physics* (New York: Dutton, 1988).

Hossenfelder, Sabine, *Lost in Math: How Beauty Leads Physics Astray* (New York: Basic Books, 2018).

Jung, Carl, *Synchronicity: An Acausal Connecting Principle*, translated by R. F. C. Hull (Princeton: Princeton University Press, 1973).

Jung, Carl, and Wolfgang Pauli, *Atom and Archetype—The Pauli/Jung Letters, 1932–1958*, edited by C. A. Meier, translated by David Roscoe (Princeton, NJ: Princeton University Press, 2001).

———, *The Interpretation of Nature and the Psyche*, translated by Priscilla Silz (New York: Pantheon Books, 1955).

Kennefick, Daniel, *No Shadow of a Doubt: The 1919 Eclipse That Confirmed Einstein's Theory of Relativity* (Princeton, NJ: Princeton University Press, 2019).

Kleinknecht, Konrad, *Einstein and Heisenberg: The Controversy over Quantum Physics* (New York: Springer, 2019).

Kragh, Helge, *Quantum Generations: A History of Physics in the Twentieth Century* (Princeton, NJ: Princeton University Press, 1999).

Kumar, Manjit, *Quantum: Einstein, Bohr, and the Great Debate About the Nature of Reality* (New York: W. W. Norton & Co., 2011).

Laurikainen, Kalervo Vihtori, *Beyond the Atom—The Philosophical Thought of Wolfgang Pauli* (New York: Springer Verlag, 1988).

Lindorff, David P., *Pauli and Jung—The Meeting of Two Great Minds* (New York: Quest Books, 2004).

Magueijo, Joao, *Faster Than the Speed of Light: The Story of a Scientific Speculation* (Cambridge, MA: Perseus, 2003).

Martinez, Alberto, *Burned Alive: Bruno, Galileo, and the Inquisition* (London: Reaktion Books, 2018).

Miller, Arthur I., *Deciphering the Cosmic Number—The Strange Friendship of Wolfgang Pauli and Carl Jung* (New York: Norton, 2009).

Orzel, Chad, *Breakfast with Einstein: The Exotic Physics of Everyday Objects* (Dallas: BenBella Books, 2018).

Peat, F. David, *Synchronicity: The Bridge Between Matter and Mind* (New York: Bantam, 1987).

Schwartz, David, *The Last Man Who Knew Everything: The Life and Times of Enrico Fermi, Father of the Nuclear Age* (New York: Basic Books, 2017).

Seeger, Raymond J., *Galileo Galilei, His Life and His Works* (London: Pergamon Press, 1966).

Smolin, Lee, *Einstein's Unfinished Revolution: The Search for What Lies Beyond the Quantum* (London: Allen Lane, 2019).

Stachel, John, *Einstein from 'B' to 'Z'* (Boston: Birkhäuser, 2002).

Stewart, Ian, *Do Dice Play God? The Mathematics of Uncertainty* (New York: Basic Books, 2019).

Strogatz, Steven, *Sync: How Order Emerges from Chaos in the Universe, Nature, and Daily Life* (New York: Hyperion, 2003).

Thirring, Walter, *Cosmic Impressions: Traces of God in the Laws of Nature*, translated by Margaret A. Schellenberg (Philadelphia: Templeton Foundation Press, 2007).

Wilbur, James B., *The Worlds of the Early Greek Philosophers* (Buffalo, NY: Prometheus Books, 1979).

Woit, Peter, *Not Even Wrong: The Failure of String Theory and the Search for Unity in Physical Law* (New York: Basic Books, 2007).

Yourgrau, Palle, *A World Without Time: The Forgotten Legacy of Gödel and Einstein* (New York: Basic Books, 2004).

Zeilinger, Anton, *Dance of the Photons: From Einstein to Quantum Teleportation* (New York: Farrar, Straus and Giroux, 2010).

REFERENCES

INTRODUCTION: Mapping Nature's Connections

1. Geoff Brumfiel, "Particles Break Light-speed Limit," *Nature*, September 22, 2011, https://www.nature.com/news/2011/110922/full/news.2011.554.html.

2. Časlav Brukner, correspondence with the author, March 11, 2019.

CHAPTER ONE: Touching the Heavens: Ancient Views of the Celestial Realm

1. The closest possibility, according to the speculations of New Zealand archaeologist Robert Hannah and his colleagues, was that a temple attributed to Juno was conceivably a center of Apollo worship because of its orientation in the direction of the constellation Delphinus, the Dolphin, associated with Delphi. See Robert Hannah, Guilia Magli, and Andrea Orlando, "Astronomy, Topography and Landscape at Akragas' Valley of the Temples," *Journal of Cultural Heritage*, vol. 25, May–June 2017, pp. 1–9.

2. Robert Hannah, Guilia Magli, and Andrea Orlando, "Astronomy, Topography and Landscape at Akragas' Valley of the Temples," *Journal of Cultural Heritage*, vol. 25, May–June 2017, pp. 1–9.

3. For example, Greek biographer Diogenes Laertius (circa third century AD) advanced that view in *Lives of Eminent Philosophers*.

4. Matthew Arnold, "Empedocles on Etna," *The Strayed Reveller: Empedocles on Etna, and Other Poems* (London: Walter Scott, 1896).

5. Daniel Chanan Matt, ed., *Zohar: Annotated & Explained* (Nashville, TN: SkyLight Paths, 2002), p. 44.

6. Aristotle, "Sense and Sensibilia," in Jonathan Barnes, ed., *Complete Works of Aristotle, Volume 1: The Revised Oxford Translation*, translated by Benjamin Jowett, (Princeton, NJ: Princeton University Press, 1984), p. 708.

CHAPTER TWO: By Jupiter, Light Lags!

1. Freeman Dyson, correspondence with the author, February 22, 2019.

2. Valery Rees, "Cicerian Echos in Marsilio Ficino," in *Cicero Refused to Die: Ciceronian Influence Through the Centuries* (Boston: Brill, 2013), p. 146.

3. Sheila Barker, "Cosimo de Medici's Chemical Medicine," *The Medici Archive Project*, March 2, 2016, http://www.medici.org/cosimo-de-medicis-chemical-medicine-2/.

4. Traditionally, Tycho's nasal prosthetic was thought to be made of precious metals such as silver and gold. However, after his remains were exhumed in 2010, chemical analysis showed that it was likely made of brass. See Megan Gannon, "Tycho Brahe Died from Pee, Not Poison," *Live Science*, November 16, 2012, https://www.livescience.com/24835-astronomer-tycho-brahe-death.html.

5. Owen Gingerich, *The Eye of Heaven: Ptolemy, Copernicus*, Kepler (New York: American Institute of Physics, 1993), p. 181.

6. Jamie James, *The Music of the Spheres: Music, Science, and the Natural Order of the Universe* (New York: Copernicus, 1995), p. 157.

7. Max Caspar, *Kepler* (Stuttgart, Germany: W. Kohlhammer Verlag, 1948), p. 117. Quoted in Arthur Koestler, *The Sleepwalkers* (New York: Macmillan, 1959), p. 304.

8. Johannes Kepler, *Astronomia Nova* (1609). Translated and quoted in Arthur Koestler, *The Sleepwalkers* (New York: Macmillan, 1959), p. 125.

9. Alberto Martinez, correspondence with the author, March 28, 2019.

10. Galileo Galilei, *Dialogues Concerning Two New Sciences*. Translated from the Italian and Latin into English by Henry Crew and Alfonso de Salvio. With an Introduction by Antonio Favaro (New York: Macmillan, 1914), p. 43.

CHAPTER THREE: Illuminations: The Complementary Visions of Newton and Maxwell

1. Isaac Newton to Richard Bentley, February 25, 1693. Reprinted in Andrew Janiak, ed., *Isaac Newton Philosophical Writings* (New York: Cambridge University Press, 2004), p. 102.

2. Thomas Bass, *The Eudaemonic Pie: The Bizarre True Story of How a Band of Physicists and Computer Wizards Took on Las Vegas* (New York: Open Road, 2016), p. 49.

3. J. J. O'Connor and E. F. Robertson, "Christiaan Huygens," University of St Andrews Mathematics Biography website, http://www-history.mcs.st-and.ac.uk/Biographies/Huygens.html.

4. Isaac Newton to Richard Bentley, December 10, 1692. Reprinted in Andrew Janiak, ed., *Isaac Newton Philosophical Writings* (New York: Cambridge University Press, 2004), p. 95.

5. Pierre Simon Laplace, *A Philosophical Essay on Probabilities (Essai philosophique sur les probabilities)*, translated by F. W. Truscott and F. L. Emory (New York: Dover, 1951), p. 4.

6. Sam Callander, "Who Was James Clerk Maxwell?," The Maxwell at Glenlair Trust website, http://www.glenlair.org.uk/.

7. Daniel Kleppner, "Master Michelson's Measurement," *Physics Today*, vol. 60, no. 8 (2007), p. 8.

8. "How Fast Does Light Travel? Experiments About to be Made to Determine the Question," *New York Times*, August 28, 1882.

CHAPTER FOUR: Barriers and Shortcuts: The Mad Landscapes of Relativity and Quantum Mechanics

1. Albert Einstein, "Autobiographical Notes," in Paul Arthur Schilpp, ed., *Albert Einstein: Philosopher-Scientist* (LaSalle, IL: Open Court, 1949), p. 10.

2. O. M. P. Bilaniuk, V. K. Deshpande, and E. C. George Sudarshan, *American Journal of Physics*, vol. 30 (1962), p. 718.

3. G. Feinberg, "Possibility of Faster-Than-Light Particles," *Physical Review*, vol. 159, no. 5 (1967), pp. 1089–1105. Note that in contemporary string theory, tachyons have a different, more technical meaning, referring to slower-than-light energy fields arising from terms with negative mass squared.

4. Albert Einstein, "Über das Relativitätsprinzip und die aus demselben gezogenen Folgerungen," *Jahrbuch der Radioaktivität und Elektronik*, vol. 4 (1907), pp. 411–462, translated and reprinted in John Stachel, David C. Cassidy, Jürgen Renn, et al., *The Collected Papers of Albert Einstein, Volume 2: The Swiss Years: Writings, 1900–1909* (Princeton, NJ: Princeton University Press), p. 252.

5. A. H. Reginald Buller, "Relativity," *Punch*, December 19, 1923 (published anonymously).

6. Gregory Benford, D. L. Book, W. A. Newcomb, "The Tachyonic Antitelephone," *Physical Review D*, vol. 2 (1970), pp. 263–265.

7. Alan Chodos, Avi Hauser, and Alan Kostelecky, "The Neutrino as a Tachyon," *Physics Letters B*, vol. 150, no. 6 (January 1985), pp. 431–435.

8. Alan Chodos, correspondence with the author, March 26, 2019.

9. Alan Chodos, correspondence with the author, March 26, 2019.

10. Joel Achenbach, "Faster-than-light Neutrino Poses the Ultimate Cosmic Brain Teaser for Physicists," *Washington Post*, November 14, 2011.

11. Deborah Netburn, "Neutrino Jokes Hit Twittersphere Faster Than the Speed of Light," *Los Angeles Times*, September 24, 2011, https://latimesblogs.latimes.com/nationnow/2011/09/faster-than-the-speed-of-light-neutrino-jokes-light-up-twittersphere.html.

12. Carlo Rubbia, quoted in Geoff Brumfiel, "Neutrinos Not Faster Than Light: ICARUS Experiment Contradicts Controversial Claim," *Nature News & Comment*, March 16, 2012, https://www.nature.com/news/neutrinos-not-faster-than-light-1.10249.

13. Alan Chodos, correspondence with the author, March 26, 2019.

14. John Stachel, *Einstein from 'B' to 'Z'* (Boston: Birkhäuser, 2002), p. 262.

15. David Wilson, *Rutherford, Simple Genius* (Cambridge, MA: MIT Press, 1983), p. 62.

16. Chaim Weizmann, *Trial and Error* (New York: Harper & Bros., 1949), p. 118.

17. Ernest Rutherford, "The Development of the Theory of Atomic Structure," in Joseph Needham and Walter Pagel, eds., *Background to Modern Science* (Cambridge, MA: Cambridge University Press, 1938), p. 68.

18. Ernest Rutherford, letter to Niels Bohr, March 20, 1913. Reprinted in Niels Bohr, *Collected Works*, vol. 2 (Amsterdam: North Holland, 1972), p. 583.

CHAPTER FIVE: The Veil of Uncertainty: Turning Away from Realism

1. Max Born to Albert Einstein, October 10, 1944. Max Born and Albert Einstein, *The Born–Einstein Letters, 1916–1955: Friendship, Politics and Physics in Uncertain Times*, translated by Irene Born (New York: Macmillan, 1971), p. 155.

2. "British Association Meets Wednesday: Sir Oliver Lodge, in Presidential Address, Will Combat the 'Theory of Relativity,'" *New York Times*, September 8, 1913.

3. "Revolution in Science. New Theory of the Universe: Newtonian Ideas Overthrown," *Times* of London, November 7, 1919.

4. "Lights All Askew in the Heavens," *New York Times*, November 10, 1919.

5. Alexander McAdie, "Alice in Wonderland as a Relativist," *New York Times*, March 11, 1923, p. 13.

6. Waldemar Kaempffert, "How to Explain the Universe: Science in a Quandary," *New York Times*, January 11, 1931.

7. William L. Laurence, "Jekyll-Hyde Mind Attributed to Man," *New York Times*, June 23, 1933.

8. Albert Einstein to Max Born, December 4, 1926. Max Born and Albert Einstein, *The Born–Einstein Letters, 1916–1955: Friendship, Politics and Physics in Uncertain Times*, translated by Irene Born (New York: Macmillan, 1971), p. 91.

9. In German, "eins" is 1 and "zwei" is 2; hence "Zweistein" follows Einstein. See John Stachel, "Einstein and 'Zweistein,'" *Einstein from 'B' to 'Z'* (Boston: Birkhäuser, 2002).

10. Albert Einstein to Arnold Sommerfeld, circa January 18, 1922.

11. David Cassidy, correspondence with the author, February 26, 2019.

12. Werner Heisenberg, interviewed by Thomas S. Kuhn and John Heilbron, American Institute of Physics Oral History, Session I, November 30, 1962.

13. Werner Heisenberg, *Physics and Philosophy: The Revolution in Modern Science* (New York: Harper and Row, 1958), pp. 71–72; https://history.aip.org/exhibits/heisenberg/p13e.htm.

14. David Cassidy, correspondence with the author, February 26, 2019.

15. Wolfgang Pauli, *Wissenschaftlicher Briefwechsel (Scientific Correspondence), Volume I, 1919–1929*, edited by A. Hermann, K. V. Meyenn, and V. F. Weisskopf (Berlin: Springer, 1979), p. 143.

16. Wolfgang Pauli, *Wissenschaftlicher Briefwechsel (Scientific Correspondence), Volume I, 1919–1929*, p. 262.

17. David Cassidy, correspondence with the author, February 26, 2019.

18. Werner Heisenberg, in S. Rozental, edited by Niels Bohr (New York: Wiley, 1967), p. 103.

CHAPTER SIX: The Power of Symmetry: Connections Beyond Causality

1. David Hilbert, anecdotal remark, reported in Nina Byer, "E. Noether's Discovery of the Deep Connection Between Symmetries and Conservation Laws," presented at "The Symposium on the Heritage of Emmy Noether in Algebra, Geometry, and Physics," Bar Ilan University, Tel Aviv, Israel, December 2–3, 1996, http://cwp.library.ucla.edu/articles/noether.asg/noether.html.

2. Albert Einstein, "The Late Emmy Noether; Professor Einstein Writes in Appreciation of a Fellow-Mathematician," *New York Times*, May 4, 1935.

3. Wolfgang Pauli, "Exclusion Principle and Quantum Mechanics," *Nobel Lecture*, Stockholm, Sweden, December 13, 1946.

4. Wolfgang Pauli to Alfred Landé, quoted in John L. Heilbron, "The Origins of the Exclusion Principle," *Historical Studies in the Physical Sciences*, vol. 13, no. 2 (1983), p. 261.

5. Ralph Kronig, interviewed by John L. Heilbron, AIP oral history interview, November 12, 1962.

6. George Uhlenbeck, interviewed by Thomas S. Kuhn, AIP oral history interview, 1962.

7. Wolfgang Pauli, "Open Letter to the Group of Radioactive People at the Gauverein Meeting in Tübingen," December 4, 1930.

8. Léon Rosenfeld, "La Plainte du Neutrino," *Journal of Jocular Physics*, vol. 1, p. 35.

9. George Gamow, reported in *Lakeland Ledger*, May 26, 1998, p. D3.

10. Barbara Lovett Cline, *The Questioners: Physicists and the Quantum Theory* (New York: Crowell, 1965), p. 143.

11. Stanley Deser, correspondence with the author, February 23, 2019.

12. Diana Kormos-Buchwald, correspondence with the author, February 21, 2019.

13. Albert Einstein, "Preface," in Upton Sinclair, *Mental Radio* (Springfield, IL: Charles Thomas Publisher, 1930).

14. C. Hartley Grattan, "Why, Dr. Einstein!," *New Republic*, March 9, 1932, https://newrepublic.com/article/119292/controversy-einsteins-endorsement-psychic-upton-sinclair-defends.

CHAPTER SEVEN: The Road to Synchronicity: The Jung–Pauli Dialogue

1. Albert Einstein to Max Born, 1952. Quoted in Flavio del Santo, "Striving for Realism, Not for Determinism: Historical Misconceptions on Einstein and Bohm," *APS News*, vol. 28, no. 5 (May 2019), p. 8.

2. Wolfgang Pauli. Reported by David Bohm, interviewed by Maurice Wilkins, September 25, 1986. American Institute of Physics Oral Histories, https://www.aip.org/history-programs/niels-bohr-library/oral-histories/32977-4.

3. Wolfgang Pauli, Review of *Ergebnisse der exakten Naturwissenschaften, 10, Band, die Naturwissenschaften* 20, pp. 186–187. Translated and reprinted by John Stachel in *Einstein from 'B' to 'Z'*, p. 544.

4. Oskar Klein. Interviewed by Thomas S. Kuhn and John L. Heilbron, Copenhagen, July 16, 1963.

5. Abraham Pais, "Glimpses of Oskar Klein as Scientist and Thinker," in Ulf Lindström, ed., *Proceedings of the Oskar Klein Centenary Symposium* (Singapore: World Scientific, 1995), p. 14.

6. Kurt Gottfried, phone conversation with the author, March 10, 2019.

7. Stanley Deser, correspondence with the author, February 23, 2019.

8. C. G. Jung, *Memories, Dreams, Reflections*, pp. 373–377.

9. Bernard D. Beitman, "Seriality vs Synchronicity: Kammerer vs Jung," *Psychology Today* blog, March 25, 2017, https://www.psychologytoday.com/us/blog/connecting-coincidence/201703/seriality-vs-synchronicity-kammerer-vs-jung.

10. Misha Shifman, ed., *Standing Together in Troubled Times: Unpublished Letters by Pauli, Einstein, Franck, and Others* (Singapore: World Scientific, 2017), p. 4.

11. Wolfgang Pauli, reported in Charles P. Enz, *Of Matter and Spirit: Selected Essays* (Singapore: World Scientific, 2009), p. 153.

12. Beverley Zabriskie, "Jung and Pauli: A Meeting of Rare Minds," in Carl Jung and Wolfgang Pauli, *Atom and Archetype—The Pauli/Jung Letters, 1932–1958*, edited by C. A. Meier, translated by David Roscoe (Princeton, NJ: Princeton University Press, 2001), p. xxvii.

13. Carl Jung to Wolfgang Pauli, October 14, 1935. Reprinted in Carl Jung and Wolfgang Pauli, *Atom and Archetype—The Pauli/Jung Letters, 1932–1958*, edited by C. A. Meier, translated by David Roscoe (Princeton, NJ: Princeton University Press, 2001), p. 13.

14. Attributed to Sigmund Freud. See, for example, Arthur Asa Berger, *Media Analysis Techniques* (Thousand Oaks, CA: Sage Publications, 2005), p. 93.

15. Carl G. Jung, *The Archetypes and the Collective Unconscious*, translated by R. F. C. Hull (London: Routledge, 1959), p. 384.

16. Carl Jung. Reported in Charles P. Enz, *No Time to Be Brief—A Scientific Biography of Wolfgang Pauli* (New York: Oxford University Press, 2002), p. 246.

17. Don Howard, "Quantum Mechanics in Context: Pascual Jordan's 1936 *Anschauliche Quantentheorie*," in Massimiliano and Jaume Navarro, eds., *Research and Pedagogy: A History of Quantum Physics Through Its Textbooks* (2013), http://edition-open-access.de/studies/2/12/.

18. Wolfgang Pauli to Carl Jung, October 26, 1934. Reprinted in Carl Jung and Wolfgang Pauli, *Atom and Archetype—The Pauli/Jung Letters, 1932–1958*, edited by C. A. Meier, translated by David Roscoe (Princeton, NJ: Princeton University Press, 2001), p. 5.

19. Martin Gardner, in Kendrick Frazier, "A Mind at Play: An Interview with Martin Gardner," *Skeptical Inquirer*, March/April 1998, pp. 37–38.

20. John R. Smythies, "Minds and Higher Dimensions," *Journal of the Society for Psychical Research*, vol. 55, no. 812 (1952), pp. 150–156.

21. Wolfgang Pauli to Pascual Jordan, March 5, 1952. Reprinted in Wolfgang Pauli, *Wissenschaftlicher Briefwechsel (Scientific Correspondence), Volume IV, Part I, 1950–1952*, edited by A. Hermann, K. V. Meyenn, and V. F. Weisskopf (Berlin: Springer, 1985), p. 568.

22. John R. Smythies, correspondence with the author, December 20, 2002.

23. Carl Jung, "On Synchronicity," in *Synchronicity: An Acausal Connecting Principle*, trans. R. F. C. Hull (Princeton, NJ: Princeton University Press, 1973), p. 114.

24. Wolfgang Pauli to Carl Jung, October 26, 1934. Reprinted in Carl Jung and Wolfgang Pauli, *Atom and Archetype—The Pauli/Jung Letters, 1932–1958*, edited by C. A. Meier, translated by David Roscoe (Princeton, NJ: Princeton University Press, 2001), p. 31.

25. Freeman Dyson, correspondence with the author, February 22, 2019.

26. Arthur I. Miller, *Deciphering the Cosmic Number: The Strange Friendship of Wolfgang Pauli and Carl Jung* (New York: Norton, 2010), p. 252.

27. Miller, p. 258.

28. Carl Jung to Wolfgang Pauli, June 20, 1950. Reprinted in Carl Jung and Wolfgang Pauli, *Atom and Archetype—The Pauli/Jung Letters, 1932–1958*, edited by C. A. Meier, translated by David Roscoe (Princeton, NJ: Princeton University Press, 2001), p. 45.

29. Wolfgang Pauli to Carl Jung, November 24, 1950. Reprinted in Carl Jung and Wolfgang Pauli, *Atom and Archetype—The Pauli/Jung Letters, 1932–1958*, edited by C. A. Meier, translated by David Roscoe (Princeton, NJ: Princeton University Press, 2001), p. 58.

30. Wolfgang Pauli to Carl Jung, December 12, 1950. Reprinted in Carl Jung and Wolfgang Pauli, *Atom and Archetype—The Pauli/Jung Letters, 1932–1958*, edited by C. A. Meier, translated by David Roscoe (Princeton, NJ: Princeton University Press, 2001), p. 64.

31. Carl Jung and Wolfgang Pauli, *The Interpretation of Nature and the Psyche*, translated by R. F. C. Hull and Priscilla Silz (New York: Pantheon Books, 1955), p. 31.

32. Reviewer's note, Carl Jung and Wolfgang Pauli, *The Interpretation of Nature and the Psyche*, translated by R. F. C. Hull and Priscilla Silz (New York: Pantheon Books, 1955); https://cds.cern.ch/record/2229568/?ln=en.

33. Wolfgang Pauli to Carl Jung, August 5, 1957. Reprinted in Carl Jung and Wolfgang Pauli, *Atom and Archetype—The Pauli/Jung Letters, 1932–1958*, edited by C. A. Meier, translated by David Roscoe (Princeton, NJ: Princeton University Press, 2001), p. 62.

34. Wolfgang Pauli to Niels Bohr, February 15, 1955, CERN (Pauli Archive). Quoted with permission from the Pauli Committee at CERN.

35. J. B. Rhine to C. G. Jung, April 24, 1959, Jung Correspondence, ETH-Zürich Archives.

36. David P. Lindorff, *Pauli and Jung—The Meeting of Two Great Minds* (New York: Quest Books, 2004), p. 238.

CHAPTER EIGHT: False Reflections: Navigating Nature's Imperfect House of Mirrors

1. Chien-Shiung Wu, "Parity Violation," in Harvey B. Newman and Thomas Ypsilantis, eds., *History of Original Ideas and Basic Discoveries in Particle Physics* (New York: Plenum, 1996), pp. 381–382.

2. Erwin Schrödinger, "Are There Quantum Jumps?," *The British Journal for the Philosophy of Science*, vol. 3, no. 10 (August 1952), pp. 109–123.

3. Jennifer Ouellette, "Madame Wu and the Holiday Experiment That Changed Physics Forever," *Gizmodo*, December 2015, gizmodo.com/madame-wu-and-the-holiday-experiment-that-changed-physi-1749319896.

4. Private communication from Georges M. Temmer to Ralph P. Hudson. Reported in Ralph P. Hudson, "Reversal of the Parity Conservation Law in Nuclear Physics," A *Century of Excellence in Measurements, Standards, and Technology*, edited by David R. Lide (Washington, DC: National Institute of Standards and Technology, 2001), p. 114.

5. David C. Cassidy, correspondence with the author, February 26, 2019.

6. Werner Heisenberg, "The Nature of Elementary Particles," *Werner Heisenberg Collected Works* (Berlin: Springer-Verlag, 1984), p. 924.

7. Stanley Deser, correspondence with the author, February 23, 2019.

8. Freeman Dyson, reported in Jeremy Bernstein, "King of the Quantum," *New York Review of Books*, September 26, 1991.

9. Niels Bohr, reported in Jesse Cohen, "Science Friction," *Los Angeles Times*, July 13, 2008, https://www.latimes.com/archives/la-xpm-2008-jul-13-bk-susskind13-story.html.

10. Quoted in David C. Cassidy, *Uncertainty: The Life and Science of Werner Heisenberg* (San Francisco: W. H. Freeman & Co., 1991), p. 542.

11. Wolfgang Pauli to Charles Enz, March 4, 1958. Translated and quoted in Charles P. Enz, *No Time to Be Brief—A Scientific Biography of Wolfgang Pauli* (New York: Oxford University Press, 2002), p. 528.

12. Wolfgang Pauli to George Gamow, March 1, 1958. Reported in Arthur I. Miller, *Deciphering the Cosmic Number: The Strange Friendship of Wolfgang Pauli and Carl Jung* (New York: Norton, 2010), p. 263.

13. Wolfgang Pauli, *Proceedings of the 1958 Annual International Conference on High Energy Physics at CERN*, edited by B. Ferretti (Geneva: CERN, 1958).

14. Charles P. Enz, *No Time to Be Brief—A Scientific Biography of Wolfgang Pauli* (New York: Oxford University Press, 2002), p. 533.

CHAPTER NINE: Reality's Rodeo: Wrangling with Entanglement, Taming Quantum Jumps, and Harnessing Wormholes

1. Freeman Dyson, correspondence with the author, February 22, 2019.

2. John Bell, reported in Jeremy Bernstein, *Quantum Profiles* (Princeton, NJ: Princeton University Press, 1991), pp. 50–51.

3. Kurt Gottfried, phone conversation with the author, March 10, 2019.

4. Anton Zeilinger, "Light for the Quantum. Entangled Photons and Their Applications: A Very Personal Perspective," *Physica Scripta*, vol. 92 (2017), p. 072501.

5. Andy Extance, "Industry Adopts Quantum Computing, Qubit by Qubit," *Chemistry World*, June 12, 2019, https://www.chemistryworld.com/news/industry-adopts-quantum-computing-qubit-by-qubit-/3010591.article.

6. Z. K. Minev et al., "To Catch and Reverse a Quantum Jump Mid-flight," *Nature*, vol. 570, June 3, 2019, https://www.nature.com/articles/s41586-019-1287-z.

7. Časlav Brukner, correspondence with the author, March 11, 2019.

CONCLUSION: UNRAVELING THE UNIVERSE'S TANGLED WEB

1. John C. Wright, "Consistency and Complexity of Response Sequences as a Function of Schedules of Noncontingent Reward," *Journal of Experimental Psychology*, vol. 63, no. 6 (1962), pp. 601–609.

2. Harold W. Hake and Ray Hyman, "Perception of the Statistical Structure of a Random Series of Binary Symbols," *Journal of Experimental Psychology*, vol. 45, no. 1 (1953), pp. 64–74.

3. Lyman Page, conversation with the author at Princeton University, April 12, 2018.

4. Časlav Brukner, "Quantum Causality," *Nature Physics*, vol. 10 (April 2014), p. 259.

5. Roy Maartens, "Brane-World Gravity," *Living Reviews in Relativity*, vol. 7, no. 7 (2004), https://arxiv.org/abs/1004.3962.

6. Wolfe Mays, "Book Review: The Roots of Coincidence. By Arthur Koestler," *Journal of British Society of Phenomenology*, vol. 4, no. 2 (1973), pp. 188–189.

7. Sting, *Lyrics* (New York: The Dial Press, 2007), p. 82.

8. Christopher Connelly, "The Police: Alone at the Top," *Rolling Stone*, March 1, 1984.

INDEX

University of the Sciences

Paul Halpern is a professor of physics at the University of the Sciences in Philadelphia, and the author of sixteen popular science books, most recently *The Quantum Labyrinth* and *Einstein's Dice and Schrödinger's Cat*. He lives near Philadelphia, Pennsylvania.